"After all, what our world is and can be are more about human imagination than, well... anything else. And isn't that a lot of what steampunk has to say? Imagine! Play! Create! Push past the artificial boundary of time to ask the real questions: what does it mean to be human? What are we going to do with all this technology? How can we create the future we want and need?"

—James H. Carrott,
Cultural Historian, 2011

The cover was illustrated by Doctor Geof

FROM THE EDITOR

It is with enormous pleasure and unbridled joy that I welcome you, dear reader, to the eighth installment of *SteamPunk Magazine*. It's been two years since we've released an issue, and a few things have changed. For starters, we exist again and will be publishing regularly! Also, I've resumed editorship. I founded this magazine, but after five issues I passed it over into the capable hands of C. Allegra Hawksmoor. Last summer, she passed it back. And we've got a new publisher. Combustion Books is a worker-run (no bosses!) genre fiction publisher based out of New York City. But while we've got a glossy cover and better distribution, we're still a committed DIY project. There will never be paid advertising in our pages. All the same people are involved. We'd like to think we've just gotten better at what we do.

If this issue is themed, it seems to be themed along the lines of mischief and criminality. We've got a rogue's lexicon of 19th-century New York City slang and the true tales of the girl gangs who ran in those streets. There's an exploration of the sky pirate archetype. There's a how-to on urban exploration (sometimes referred to as "trespassing") and DIY instructions for growing poppies to make one's own laudanum. These latter and potentially illegal topics, of course, are intended only to inspire our fiction and our imaginations.

Social unrest and protest are on everyone's tongues as much now at the start of 2012 as they were in the 1870s, and we would be remiss if we did not include such topics in our pages. We've an interview with activist cosplayers Steampunk Emma Goldman and Voltairine DeCleyre, who introduce us to the ideas of anachro-anarcho-feminism. We've a brilliant essay on Occupy and steampunk. We've also taken a look at the Haitian anti-colonial revolution, since anti-colonialism is dear to our little brass hearts!

Not everything is drugs and revolution, of course. There're some mathematical proofs to get you started dreaming of lighter-than-air craft. There's an introduction to those internally-combusting dieselpunk kin of ours. There's a simple and beautiful sewing project, expertly explained. We've got interviews with a maker, a cellist, a comic writer and a smut writer. We've spoken with those who run *Ruggine*, an Italian steampunk fiction journal.

And fiction. Oh so much fiction. Our fiction will take you from the American Civil War to the immigrant tunnelers who built New York City's subway system; it will take you into madness and into Eastern Europe. It will take you beneath a fantastical city run by a machine and far into the post-apocalypse.

It's here in the letter from the editor that I get to say whatever I want, so what I want to say is this: steampunk has re-inspired me. I don't know if it's a good idea to let people know that I ever doubted steampunk. I'm probably supposed to lie and claim I've always thought it's perfect in every way. But the reason I put down the magazine three years ago is that I was burned out. I still loved steam engines and mad science, I still read more 19th-century political theory than 20th. I still wrote about crazy contraptions and I still wore entirely too many earth tones to fit in with the punks that I ran with in my day-to-day life. But I'd grown tired of the culture. I got into steampunk with naive assumptions. I thought it would be all tea and regicide, all the time, but it wasn't. Sometimes—and haven't we all felt this way?—steampunk was just some gears hot-glued onto the boring mainstream culture I'd long ago rejected. I took my leave and retired to my books and the aforementioned tea and the company of those who shared my passion for a life less common.

But then, last summer, I found myself attending the Steampunk World's Fair in Trenton, New Jersey. Circus punks marched through the hallways with drums and saxophones while workshops covered doing steampunk cheap and doing it anti-colonial. A labor rally in the courtyard brought our 19th-century flair to 21st-century problems. Academics presented literary criticism—and if I recall, there was a drag show. People from every generation attended, and everyone I met was amazing. I was humbled.

I love this culture even if it's different from what I expected it to be. And coming back to the crew of *SteamPunk Magazine* felt a lot like coming home. For a full-time nomad, that's saying something. Steampunk, you have never failed me. You are awesome because the people make you awesome. I'm glad to be back.

—*Margaret Killjoy*

CONTENTS

What is Steampunk? What is SteamPunk Magazine?

The term "steampunk" was coined to refer to a branch of "cyberpunk" fiction that concerned itself with Victorian-era technology. However, steampunk is more than that. Steampunk is a burgeoning subculture that pays respect to the visceral nature of antiquated technology.

It's about "steam," as in steam engines, and it's about "punk," as in counter-culture. For an excellent manifesto, refer to the first article in our first issue, "What Then, is Steampunk?"

SteamPunk Magazine is a print publication that comes out erratically. Full quality print PDFs of it are available for free download from our website, and we keep the cost of the print magazine as low as possible. Back issues can be downloaded and the first 7 issues have been anthologized in a reader available from our publisher, Combustion Books. All work on the magazine, including articles, editing, illustration, and layout, is done by volunteers and contributors. To see how you can get involved, see page 109.

WWW.STEAMPUNKMAGAZINE.COM
COLLECTIVE@STEAMPUNKMAGAZINE.COM

RIOT GRRLS 19TH-CENTURY STYLE

By Screaming Mathilda
Illustration by Juan Navarro

The Victorian Era's suffragettes and bluestockings are abundantly recognized for their struggles to secure women's rights and for spawning the political and philosophical basis of modern feminism. While these upper- and middle-class women were seeking to reform the relations between the sexes primarily within their class, the myth that lower class women in Victorian times did not seek or fight for freedom—and mostly relied on the charitable efforts of their upper class benefactors—was born. It is curious that most of us have heard about the suffragettes and the settlement houses, but we know virtually nothing of the struggles of lower-class women. These women were already demonstrating that indeed sisterhood is powerful while their upper-class sisters were politely petitioning for reform. The fighting lower-class women, and sometimes girls, of New York City were not giving speeches in convention halls, they were not rubbing shoulders with city officials in the corridors of power, they were not holding soirées in Gramercy parlors; they were living radical feminism on the mean streets of Hell's Kitchen, the Five Points, and the Lower East Side.

Despite its awesome fashions, the Victorian Era was a time when it sucked to be a woman even in the upper and middle classes, and it sucked even harder to be a poor woman. In the teeming impoverished neighborhoods of Lower Manhattan, survival, for independent-minded women, often meant living outside the law. Gangs were a road to status in those neighborhoods, as they are in many poor urban areas today. When we think of gangs in New York during this period we think of toughs in dented derbies. The photos of Jakob Riis and the film *Gangs of New York* by Scorsese come to mind. In modern literature, the women of this era are either refined, high-class ladies or desolate prostitutes. Yet there were other women whose history has almost been lost to us, women with strength and courage who were able to fight sexism on the streets and to achieve amazing things.

6'6"
6'
5'6"
4'
NAV
2011

Women were a major part of gangs in the 19th century because there were so few other opportunities available to poor, young, urban women of the time. This was not only true in New York but all major cities, including Chicago, London, Paris, and Manchester. Poor men had more job options and some even had access to limited educational opportunities, but women were limited to a few poorly-paid menial jobs, prostitution, or child-rearing. For a girl that did not want to marry, there were few legitimate avenues open. Gangs opened up a path for self-sufficiency, money, and status in the lower-class neighborhoods of large cities.

There were a number of prominent female gang leaders that rose to the top of mostly male gangs in New York. Most of these women have been reduced by history to footnotes while male gang leaders' stories are much better documented. And these are only the women that somehow made it through history to us—there were many more whose names we will never know. What follow are some of the most interesting female gang leaders of New York during the Victorian Era.

One of the most famous was Hell-Cat Maggie, a poor Irish girl from a large family in the notorious Five Points area of Lower Manhattan. As a little girl, she had the odious job of matchstick girl. Matchstick girls had to pay for the matches, dip them in phosphorus, and then sell them on the street. Many matchstick girls were suffering from hair loss and phossy jaw (a deadly form of bone cancer that could turn a whole side of a person's face green and then black) from the constant contact with phosphorus. At the tender of age of eight, Maggie was selling matchsticks when she came across the sneak thief Gentleman Jasper. Jasper was trading blows with two street toughs in an alley behind the city morgue. When one of the boys stopped his assault on Jasper and tried to take her earnings, instead of meekly giving them to him, Maggie attacked him with her nails and teeth, and the young tough fled. Jasper gave her the name Hell-Cat after that fight and introduced her to his gang, the Whyos. The Whyos, one of the biggest gangs in the area with a few hundred members, were so-called because the gang's cry sounded like a bird call. Maggie was surprised to find many girls in the gang and soon joined, giving up match-selling forever. She learned the art of sneak thievery from Jasper and would often accompany him as a decoy. She was a fast learner and had an eye for detail, and soon she was the best sneak thief in the Whyos.

But when the Whyos started to work for the Know-Nothings (a Nativist political party that hated Irish immigrants) as "shoulder-punchers," forcing people to vote for a particular political party, Maggie left the Whyos and joined the famous Irish gang the Dead Rabbits. She became known as a fearsome street brawler, filing her teeth to points and using sharpened brass tips on her fingers as claws. She only fought anti-immigrant gangs and cops. Even though she was part of the Dead Rabbits, Maggie was known as independent and would work and fight with many gangs in the area. She remained partners with Jasper but refused to marry him. He eventually died from injuries received in an ambush by another gang. Maggie then started only wearing black and gave most of her thieving proceeds to other neighborhood widows. In some accounts, Hell-Cat Maggie is confused with another famous female brawler called Gallus Mag.

Gallus Mag, the six-foot bouncer of a Water Street joint called the Hole-in-the-Wall (which had two other female bouncers), would drag troublemakers out by one ear held in her clenched teeth. Mag got her nickname "Gallus" because she wore men's suspenders ("gallus" as they were called at the time) on her skirt. The Hole-In-The-Wall was one of the few bars where gays and straights could mingle in New York at the time, and it was rumored that Mag was a lesbian, but no one ever questioned her about it. She stalked the basement tavern looking for troublemakers with a pistol stuck in her belt and a bludgeon strapped to her wrist. If someone was dumb enough to cause any trouble or make a sexist remark, she would hit him with her club, clutch his ear in her teeth, drag him to the front door, and throw him out into the gutter. If the guy put up a stink, she would bite his ear off and store it in a large bottle of alcohol she kept in plain sight behind the bar labeled "Queen Victoria Eggs." The New York City police had proclaimed Gallus Mag the most savage female they had ever encountered, but they never dared to arrest her.

Another famous female gang leader was Sadie Farrell, a.k.a. Sadie the Goat. Little is known about Sadie's early life. It is reported she was raped as a young girl by a prominent neighborhood gang member. The story goes that the next day she borrowed knives from her brother, who was a butcher's apprentice, and took revenge on the man who had violated her. Some say she cut off his penis, others say she slit his throat. She then floated among a few gangs in Hell's Kitchen before joining the all-male Daybreak Boys. She forced them to change their name to the Daybreak Boys and Girls, even though she was the only woman in the gang. She gained a reputation for robbing rich East-Siders by first head-butting them in the stomach. In the 1860s she joined the Charlton Street Gang, river pirates on the West Side. According to Herbert Asbury's 1928 book *Gangs of New York*, Sadie got her nickname from her habit of chatting up a likely mark, then butting him in the stomach. "Under her inspired leadership, the Charlton Street thugs considerably enlarged their field of operations," Asbury wrote. "They stole a small sloop of excellent sailing qualities, and with the Jolly Roger flying from the masthead and Sadie the Goat pacing the deck in proud command, they sailed up and down the Hudson from the Harlem River to Poughkeepsie and beyond, robbing farmhouses and riverfront mansions, terrorizing the hamlets and occasionally holding men, women and children for ransom." It was during this period that Sadie, full of money and confidence, walked into a bar in the Fourth Ward. She got into it with a patron who tried to grab her breast. There was a fight, and the local bouncer, Gallus Mag, tried to clear the bar. Everyone left except Sadie, who demanded the right to finish her drink since she had been rudely interrupted by the sexual assaulter. Gallus and Sadie exchanged words, then traded blows, and in the end Sadie left without one of her ears. Sadie the Goat was said to have returned to the Fourth Ward a year later and paid her respects to Gallus Mag. She got her ear back as a token of peace. The legend says that Sadie was so overjoyed by the return of her ear that she wore it in a locket around her neck until her final days. It also goes that Gallus and Sadie moved in together and were seen together until Gallus mysterious death at the Bucket of Blood bar.

Ida Burger, called Ida the Goose, started out as a prostitute. She had a pimp that stole from her, so she bludgeoned him to death with his own walking stick on a crowded city street. She was arrested and spent some time in jail. When she returned from upstate she joined the Gophers gang (they got the name from their cellar hideouts) and immediately started a Ladies Gophers of which she was the head. The Lady Gophers were also called the Kissing Ladies because most of the women were supposedly gay. At the turn of century, she was lured away to the Lower East Side's Eastman Gang, led by the infamous Monk Eastman, but eventually went back to the Gophers after a bloody shootout. Ida's claim to fame was a fashion trend she started. Looking for a unique gift for her girlfriend, she blackjacked a police officer and stole his jacket. Her girlfriend quickly stitched it into a "smart jacket of military cut" and created a new, short-lived, fad. After several officers met with the same fate, police began patrolling Gopher territory in groups. Ida once told a reporter: "Most girls are victims. It's society's fault, not the girls', but give a girl an education she can be a banker and give her a knife she can be a gang leader. It was easier to get a knife in my neighborhood." When she died, men from various gangs she worked with wore her trademark green skirt for a week to show their respect.

Hell-Cat Maggie, Gallus Mag, Sadie the Goat, and Ida the Goose were not the only women to break the myth of the fairer sex. There were whole gangs of anonymous women that were organizing for mutual aid and self-defense against men and society as a whole. What is surprising is that these women-only gangs often lasted much longer than their male counterparts. British girl gangs like the Forty Elephants and the Badgers numbered in the hundreds and lasted for over 60 years. The same was true of many of the women-only gangs of New York. The most famous women's gang in New York lasted 35 years and was called the Battle Row Ladies' Social & Athletic Club. The police recorded that the BRLS&A would swarm a group of legitimate laborers (e.g. strike-breakers or scabs) and provide female muscle by "biting and scratching" while girls as young as five would throw bricks from the adjacent buildings down on the heads of the police to keep them away. Once the BRLS&A actually marched into a police station where one of their own was being held and proceeded to beat the cops and release their comrade. The BRLS&A would also provide "safety" committees during the frequent city riots. These safety committees would protect women from being robbed and raped by drunken looters and rioters. The ladies fought successfully to protect a female workhouse during the draft riots. Other female gangs in New York at the time included the Lady Gophers, 4th Street Amazons, The Merry Mollies, the Lady Shirttails, the Girls Roach Guard, the Slap-Dashers (also known as the Slip Dashers) and the Daughters of Erin. Many of these gangs would engage in group actions called "panics." A panic was when the gang's girls would descend en masse on a department store and ransack it, using numerous entrances to confuse staff and pilfering whatever they wanted before fleeing. If the girls were ever caught, police knew they would put up a ferocious struggle. A favorite fighting technique was covering ones fingers with diamond rings and punching police in the face and eyes. Police often sat helplessly watching a hundred women loot fashionable stores on Broadway. Panics became so commonplace that in 1882 shopkeepers could buy "panic insurance" for their wares. Queen Anne, one of the gang leaders, was reported as saying in court, "We fight scabs because you kill our men if they do. We take things [from stores in panics] because we have no money for them. You jail us because you will not let us earn our daily bread."

The poor women and girls that turned to gangs in the 19th century will probably never be allowed to take their seats in the feminist pantheon along Susan B. Anthony, Jane Addams, or the other prominent 19th-century women's rights advocates. Yet they were fighting for related ends, just in a different place and situation. Sadie the Goat and the Battle Row ladies didn't have the luxury to wait for opportunities to be given to them—they had to make their own luck where they could and with what they had.

ROGUES LEXICON:

New York Victorian Street Slang

by Professor Calamity
Illustration by Dr. Geof

It is not uncommon to find steampunks today using old-fashioned words and phrases when they're speaking and writing. Unfortunately, the majority of anachronistic words come from the sterile bell-jar of upper-class British society. This is understandable of course, since much of the Victorian literature that survives today was written in this style. Even Charles Dickens did not use contemporary street slang in his stories. It was during the Victorian Age that many intellectuals became convinced that the "pure" English language was being polluted by the poor and immigrants from the colonies.

They were publishing guidelines for everything from newspapers to textbooks to advertising slogans that restricted the growth of English. Despite these laws and regulations, the English language continued to evolve and change. Nowhere was this more obvious than in New York City. The American Victorian period saw an explosion of street slang that influences formal and informal English to this day. Words like *cool, hip, pal, booze, bummer, lush,* and phrases like *to cap someone* or *drink like a fish* all date back to this remarkable period.

So where can the intrepid steampunk go to find authentic Victorian street slang from the time before mass recordings? A particularly fruitful place to search is in the "Rogue Lexicons" created by various police forces during the Victorian period in America's big cities. In 1859 New York City's police commissioner, George Matsell, published the first cant dictionary in America, entitled *Vocabulum or The Rogue's Lexicon.* More good hunting grounds for obscure and forgotten slang are the pages of dime novels (the US equivalent of the British Penny Dreadfuls), such as Deadwood Dick. There are also some copies of "block papers" in microfilm at various libraries. The block papers were local newspapers that often only consisted of a few pirated articles from other regular dailies and a local crime blotter.

For those of you who are too sensitive to read the suspenseful and gory yarns of the dime novels or don't have the eyesight for microfiche, we have provided a mere sampling of some of the colorful street slang of Victorian New York.

ADAM: pal
Molly isn't my girlfriend, she's my Adam.

AMBIDEXTER: wishy-washy; takes both sides
When Lefty crafts a scheme he's no ambidexter.

AMBUSH: fraudulent weights used by grocers
Beware the butcher on Mulberry. He's got an ambush under the counter.

BAPTIZED: watered-down alcohol
The whiskey at McGurk's is always baptized.

BETTING HIS EYES: cowardly; overly cautious
Ever since Lefty got out of jail he is only betting with his eyes.

BINGO: alcohol. A "Bingo-boy" is a drunk.
Lefty used to be a bingo-boy, spending all of his money on bingo.

BLACKLEG: gambler
Lefty likes to go to Las Vegas to swindle the other blacklegs.

BLEAK: handsome
Lefty's new hat is bleak.

BLUDGET: first meant "female pickpocket" but later came to mean a free-spirited female
Scarlett Johansson started off in film mostly playing bludgets.

BLUNT: money
Molly has all the blunt.

BUTTERED: beaten
Those stupid skinheads got buttered by those punks.

CHUMP: head
It is hard to find a hat to fit my chump.

CHEESE IT: shut up
Sometimes it is better to just cheese it when the cops are around.

CHOCKER: clergy
Only talk to your lawyer or a chocker.

CLEAN: smart
That detective was clean, he saw right through the pair's ruse.

CROAKERS: newspapers
Never believe what you read in the croakers.

DEVIL'S SPIT: worthless
The New York Post is devil's spit in my opinion.

DICK: swell; cool
Molly's new skirt is so dick!

DIP: two cents
My boss wouldn't give me a dip if I needed it.

FLAG: five bucks
Lefty will sell you a bag for a flag.

FLASH: slang
I could barely understand Molly because she peppered everything with flash.

FLASHY: someone who knows the street; streetwise
If you're starting out, there's no better teacher than Lefty. He is so flashy.

GANDER: someone who is unfaithful; a player
I don't know what Molly was thinking. Lefty has always been a gander.

GETTING BONED: arrested or caught
After that robbery it was only a matter of time before Lefty got boned.

HAYSTICK: cigarette. "Hay" is tobacco.
Hay will kill you; that's why I gave up those haysticks a few years ago.

HOPPER: dope fiend; junkie
A couple of hoppers broke into the pharmacy last night and stole everything.

LAMPS: eyes
This guy had the most beautiful lamps I have ever seen.

NANCIES: male homosexuals
I was at the gym working out with a couple of super-buff nancies.

NOSE: informer or snitch
Lefty is a lot of things but he's not a nose.

PEEPS: children
Sally has more peeps than names for them.

PIGEON: victim of a crime
The pigeon picked Lefty out of the line-up.

PUSH: power
Lefty's got push around here.

QUEEN DICK: it never happened.
As far as you know that robbery is Queen Dick.

ROOKERY: ghetto
I wouldn't go to the rookery unless you got friends there.

REAM: cool
That new club on Houston is ream.

SALT BOX: jail
Lefty did five years in the salt box for robbery.

SAND: guts; balls
Molly got more sand than Lefty, that's for sure.

SLAP-BANG: poorly done
Lefty could have gotten away with it if the plan wasn't so slap-bang.

SOFT: paper money
I got pockets full of soft and I'm looking for a place to spend it.

STAR-GAZER: prostitute
Molly works nights but she is not a star-gazer.

TAKE THE SHINE OUT: cut down someone/something
Molly knows how to take the shine out of her uppity co-worker.

TOFF: well-dressed
Lefty is one toff dude.

TWIST: girl
I got to get some flowers for my twist.

WHISKERS: counterfeit
Those Versace handbags in Chinatown are all whiskers.

WOODEN COAT: coffin
Lefty was going to end up wearing a wooden coat if he didn't pay Molly back.

YACKS: cops
Never trust the yacks.

WHY STEAMPUNK (STILL) MATTERS

"There is nothing better than imagining other worlds ... to forget the painful one we live in. At least so I thought then. I hadn't yet realized that, imagining other worlds, you end up changing this one."
—Umberto Eco, Baudolino

by James Schafer & Kate Franklin of PARLIAMENTANDWAKE.COM

WE HELP ADMINISTER ONE OF THE LARGEST (VIRTUAL) communities of self-identified steampunks, Steampunk Facebook. There isn't a reliable way to assess the opinions of our one hundred thousand members, but from our entirely subjective assessment of the community we have become increasingly convinced that as a movement of social revolution Steampunk has failed. To be fair, there were those who argued, sometimes quite vehemently, that no such revolutionary program had ever begun—but we were some of the few who wanted to believe. We were never convinced that people were only attracted to steampunk because it looked cool and made a great setting for adventure novels and RPGs. Instead we believed that steampunk's appeal was its inherent rejection of disposable consumerist culture and the dominance of our contemporary society by modern day robber barons. We felt that, even if most people couldn't enunciate it, they were embracing steampunk as a way to deal with the pervasive unease experienced by nearly everyone raised in the West on a steady diet of ideas like "planned obsolescence" and "for-profit health care"—on ideas spawned by a nineteenth century capitalist ethos run amok with twenty-first century technology. Frankly, we still believe that. Unfortunately, we can't deny the reality that this hasn't created a community of steampunks who seriously adhere to a revolutionary, or even a particularly progressive, philosophy. There are certainly still steampunks out there fighting the good fight and there are significant overlaps between steampunk and the aesthetic tastes of many social experimenters and counter-culture artists—but that's not the same thing as saying that steampunks as an "interest group" seriously endorse any progressive real world agenda.

It's not even clear what the current significance is of the anti-authoritarian literature (Michael Moorcock's *Warlord of the Air*, William Gibson and Bruce Sterling's *The Difference Engine*, etc.) which drew many of the old guard of steampunks to the movement with its irreverence for aristocrats, industrialists, militarism, imperialism, and crass commercialism. To the extent that steampunk remains a literary culture at all, most of the books in question have the feel of romances and pulp adventures (Gail Carriger's Parasol Protectorate books, Geoff Falksen's Hellfire Chronicles, Chris Wooding's *Retribution Falls*, etc.)—entertaining, but far more ambiguous in their social commentary. And, to be frank, it seems like just as many or more of the youngbloods are here for the costumes (which are as likely to be aristocrats, industrialists, military policemen, and imperialists as revolutionaries and pirates) as for books of any stripe. This trend is only accentuated by the relentless efforts of retailers both independent and multinational to cash in on the appeal of steampunk. In short, when Justin Beiber's handlers are dressing him in a steampunk costume we can be confident that if steampunk ever had claws with which to scare the establishment, they have since been removed.

These realizations have forced us to do a lot of soul-searching. Our own aesthetic tastes are probably closer to steampunk than to any other style tribe, but is that enough to justify the hours we spend every day producing and narrating content for steampunk as a community? Does any of this matter? Is anything that we do making the world a better place, or would we be better off campaigning for local political candidates or, for that matter, just reading and playing video games? We came up with some unexpected answers to those questions that we're going to share in the next few pages—but the précis is: yes, steampunk still matters because it allows us to imagine change, and that is the most important step in ultimately making such change a reality.

By serendipity, we encountered, in rapid succession, a similar idea twice from sources at very different places on the political spectrum. First, Mark Stevenson in his *An Optimist's Tour of the Future* suggested that our contemporary society has been crushed by cynical dystopian views of the future. He argued, quite eloquently, that if every vision we have of the future is dismal we're guaranteed to live in such a future. We might not have flying cars even if we imagine a future with them, but if we can't imagine such a future, we're certain not to realize it. He is a wry man, but he appears to make his living giving motivational speeches to executives and generally arguing that the whole system of oppressive corporatist rule will work out its kinks. At the other end of the spectrum is David Graeber who was released from Yale for union organizing and who was instrumental in making the Occupy Wall Street (OWS) movement happen. Graeber argues that the most important victory of global capitalism in the last few decades hasn't been material, but psychological in that it has robbed us of the ability to imagine a world in which the corporate plutocrats aren't our overlords. Likewise, he argues, to cast off their rule the first thing we have to do is imagine a world, or many different worlds, where we do things differently. These authors are both drawing on older, more elegantly simple arguments from philosophers of history and science like Michel Foucault, who demonstrated that the history of western science (of which science fiction, be it tales of man-eating manticores or airships, has always been a part) is not a unilinear narrative of ever-increasing rightness, but rather a series of violent shifts in what is imagined to be possible. The message for a socially motivated imagination is the same: if we're unhappy with the way things are, or with the place we think society is heading, we have to visualize some alternate destination.

This isn't all the business of rarified theory; the real world is redolent with instances of imagination shaping existence. From a technological perspective, we only create what we believe to be possible. Morse created an effective telegraph system and the Wright brothers took flight in the Kitty Hawk not as isolated mad inventors, but amongst of a sea of competitors—all of whom had succumbed to a zeitgeist that believed such inventions could happen. Sometimes that knowledge is even more concrete, as in one of our favorite examples: the alphabet created by the Cherokee people. These innovators could not read European languages, but they had come in contact with Europeans using writing and so they knew that such technology was possible. Inspired, they invented their own. Time machines, faster-than-light travel, and machine consciousness may be as beyond us today as telephones, submarines, and aeroplanes were beyond da Vinci; but if we don't consider the possibility of such technologies were are certain not to create them.

Social and political history are rife with comparable examples. It was once certain that women/non-whites would never be able to hold political office/practice medicine/be soldiers/etc. and then someone suggested that maybe they could. Then a lot of people considered the possibility. Then it went from being a possibility to an experiment. Then it became real. In the weeks before OWS the dominant political buzzwords were debt and deficit, in the weeks after they were jobs and unemployment. It remains to be seen whether that change in dialogue, in the imagined possible, will result in concrete change in policy. However, if we as a society aren't even talking about job creation and the corporate corruption of government, we certainly aren't going to address those issues. We believe that OWS has been so popular primarily because it has allowed a generation that had known a decade of despair and hopelessness to see that we *could* have different agendas and *could* have a world with a different distribution of power.

Sadly, such imagination to change can work in negative directions as well. If you'd asked the best minds of the early nineteenth century where they saw their civilization headed they'd have talked about cosmopolitanism, internationalism, brotherhood based on common interests, anti-clericalism, and the triumph of reason. By the end of the nineteenth century, for a bevy of complex reasons, a completely different set of ideologies had become dominant. Brilliant people imagined the world as carved up into zones of control based on ethnic empires. The conviction that the most important distinction between people was their national/racial origins was so strong that when World War One arrived even the European socialists, men and women who had preached an international fraternity of labor, voted (with a few exceptions) for war. One could argue that their failure to resist the drive to war was a failure of vision—a failure to see the world in terms other than a life-and-death struggle between largely manufactured and mythological nationalist identities, and their co-constructed "enemies."

Science fiction has long had a role in shaping the popular imagination of the possible, including both desirable paths and those we would reject.

In most of the cases outlined above the transition from imagination to reality has been amorphous and indirect. But there is also a long tradition of more explicit programmatic arguments dating back well before Thomas More's *Utopia* and encompassing movements as diverse as the various real and imagined communes of the eighteenth and nineteenth centuries (Coleridge's Pantisocracy, Hawthorn's Brook Farm, etc.), the hippie movement of the mid twentieth, internal revolutions (the Russian being the most notable) and those against colonial rule (American, Indian, Haitian, etc.). Such utopian programs need not be successful in realizing their aims: the imagined world of Marx that fueled the October Revolution and began in such hope, for example, ended with the bureaucratic totalitarian tyranny of the Soviet Union. But they are not always failures, either. It is difficult to imagine the overthrow of aristocratic rule in late eighteenth century France without the preceding generation of Enlightenment philosophers who had posited the vision of a republic with universal human rights.

Science fiction has long had a role in shaping the popular imagination of the possible, including both desirable paths and those we would reject. Jules Verne, the proto-steampunk saint, is decidedly in the former camp and is frequently credited with inspiring most of the great technological inventions of the twentieth century. But he was just the most prominent of a large body of nineteenth century futurists who speculated on the machines lurking just over the temporal horizon (see, for example, Ashley's *Steampunk Prime* and Clarke's *The Tale of the Next Great War* as well as Jess Nevins' excellent series of "Victorian Hugos"). In most cases it's probably not possible to establish a causal link between any of these writings and the research done to make something like their visions viable; but it's easy to imagine them contributing to a general belief that a particular invention would be possible. In contrast, while Verne seemed most concerned with the technology itself, H.G. Wells, polymath that he was, was more interested in the interface of that technology with civilization and in using fantasy as a tool for writing fables about today. The same can be said of E.M. Forster, who described himself as "a liberal … who found liberalism crumbling beneath him," and whose short story *The Machine Stops* belongs on everyone's steampunk reading list. Their dystopian visions functioned more as warnings and as contemporary social commentary than as roadmaps, and along those lines we're reminded of George Orwell, who wrote in *1984*, "If you want a picture of the future, imagine a boot stamping on a human face—forever." There has perhaps never been a more effective conjuration of a possible world, nor a more strident call for us to ensure it never comes to be, than Orwell's haunting vision. Writers are products of their times, so it's no surprise that the twentieth century gave birth to as many science fiction dystopias as utopias; unfortunately, one can only go so far by listing all the worlds we don't want to create and crafting dystopias to reveal the nightmares lurking in contemporary political visions. If science fiction is going to help make the world better it must do more than terrify us with visions of apocalypse—it must balance the dystopian and the utopian. With the advent of the god-like computers of the post-Singularity genre, optimistic visions of the future have come back into fashion; but such fiction offers little in the way of advice beyond "wait eagerly for the computer scientists to build the first artificial intelligence." Praying for machine saviors isn't any more helpful than utopian dreams built on the crumbling foundations of our current system. Where are the visions of a different, better, future that could be built by people like us? Our answer, of course, is that they are—or at least that they could be—found in steampunk.

While on first inspection steampunk is obsessed with its gadgets and its fashion, at its heart it is a genre about people and society—and people and society recognizably related to our own. It is not (primarily, at least) a genre of intergalactic empires, fairy-witch-princesses, or near-omnipotent-alien- or artificial-intelligences; but rather it is one usually focused on people coping with technological revolutions and social realignments within worlds possessed of significant wealth and power asymmetries. While pirates, analogous to their hacker cyberpunk precursors, are much loved anti-establishment protagonists, the action more frequently centers around comparatively unremarkable laborers (see Dexter Palmer's *The Dream of Perpetual*

Motion and Cherie Priest's *Boneshaker*); and when society's elites like scientists (*The Difference Engine*) and aristocrats (Mark Hodder's *The Strange Affair of Spring Heeled Jack*) are the focus, they are frequently depicted as wrong-footed in worlds made unstable by strange technology. While there may be airships and rayguns, even magic and monsters, the worlds and their inhabitants are familiar enough that we could imagine their world as our own or—perhaps more importantly—imagine us remaking our world to be theirs. This is why it is so powerful when authors push the boundaries of the socially "possible," as they do in two of our favorite Steampunk novels.

Neal Stephenson's *Diamond Age* is set in a near future world in which nanotechnological advances both motivate and enable a tiny elite group of steampunks to create their vision of a neo-Victorian world, complete with a new Queen Victoria and exploitative colonial relationships both with their Asian neighbors and the client communities that produce the handmade luxury goods they value. At first glance the novel appears to be an ode to these "Vickys" and their innovative lifestyle; and there is no question that if "we" could ensure that our future involved "us" being the Vickys and not their subjects it would be a wonderful world to emulate. However, we would argue that the more compelling parts of the novel take place in the peripheries produced by the Vicky consumer empire. In particular, in the social turmoil of a resurgent China where an effort is underway to appropriate the means of production away from elite chokepoints and to transfer it, potentially with disastrous results for the current social order, to everyone; and in the effort of one of the iconoclastic founders of the Vicky world to create an educational system that would train the next generation to be both responsible to the community and revolutionary. It is in those descriptions that we see an alternate model for a new world of "makers" and innovators.

China Mieville in his Bas-Lag trilogy, particularly in the second (*The Scar*) and third (*Iron Council*) books, shows us entirely alternate societies. In *The Scar* it is a loose republic of pirates who have built an amalgamated society with a sociopolitical culture that mirrors the artificial island of diverse ships on which they live. In *Iron Council* it is a revolutionary commune created by striking railroad workers, the sex workers previously used to keep them docile, and the slaves intended as scabs, all of whom together build a mobile society that lives off an appropriated train. (If neither one of these captures your interest he also throws in a staggeringly diverse assortment of matriarchies, kleptocracies, hive-minds, and more conventional tyrannies.) Mieville is more self-consciously prescriptive in his writing than Stephenson, and you can almost feel him testing out each of his model societies for its potential as a roadmap to real world social transformation. As you read about each one you find yourself doing the same thing—most you initially reject out of hand, but if you let yourself, you'll return to them and end up asking whether it might not just be possible. Could it happen here? Could there be a world in which all the power didn't belong to a few corrupt elites?

These two novels have in common the fact that they aren't alternate histories. Using Falksen's popular but ultimately procrustean definition that "Steampunk is Victorian Science Fiction" one certainly can conjure alternate worlds (Moorcock, Gibson, and Sterling did just that when they gave birth to modern Steampunk and Cherie Priest, Scott Westerfeld and a host of others have done it with varying degrees of success subsequently), but to some extent those worlds will be hidebound in the social contradictions of the past. This makes them very good for problematizing our Victorian history (and thus our present), as some of these authors did with novels that attacked social injustice and imperialism; but they have a much harder time offering up alternatives that could serve as desirable models unless they rewrite history so extensively that they are really no different from novels set in some altogether alternate, if bustle-loving, universe. Similarly, the revision of history with the inclusion of popular modern subjectivities like the "cantankerously-emancipated woman" trope and modern or post-modern representations of class, race, nationality, or gender leads to narratives that may critique our history and present using a contemporary morality, but which also struggle to avoid reinforcing current constructions of society and self. This, of course, is a spectrum as alternate history bleeds into the fantastical, and it's conceivable that someone could write a historical novel that questioned not just historical but present constructions of identity in the same way that Ursula K LeGuin's classic science fiction work, *The Left Hand of Darkness*, questioned fundamental assumptions about the very nature of gender in society. Our salient point is that while the application of the steampunk aesthetic to history is an inspiring means to understand and critique our present, we believe applying that dissatisfied, innovative aesthetic to Brave New Worlds is a better way to visualize and thus achieve alternate futures.

We have mentioned some of the very best that steampunk fiction has to offer—in terms of quality they are the exception. While Charlie Stross has gotten the most attention for his critique of steampunk as bad genre fiction, he is not the only person to feel frustration at the proliferation of titles with gears on their covers and unreadable prose within. In the past year we've had to put down almost as many steampunk novels as we've finished since much of the stuff being pushed out is bad, sometimes downright terrible. We believe, however, that this misses the point that bad genre fiction is, more importantly, also just bad fiction. We can't expect steampunk novels (or movies, or art, music, etc.)

to be better than vampire romance or space opera or any other broad category that has excited people (and publishers) and which thus is generating a lot of content. People are enthusiastic about the genre so there is more of it being written—and the fact that much of it is derivative and forgettable doesn't make it any different than most of the rest of the stuff at the bookstore. We're still reading Jane Austen and Emile Zola while most of their contemporaries (and they had many contemporaries) are forgotten—it's unfair to expect anything different from the popular literature of today. But rather than despair, we believe that all that crap steampunk fiction can do useful work, particularly if its authors look to the right places for inspiration. In the age of Twitter and the 24-hour news cycle, one can't make a point simply by saying something smart and important. One has to get other people repeating it, making it more accessible to other audiences, repackaging it, diluting it, passing it back and forth, causing people to hear it from multiple sources at once. This kind of media echo chamber can be incredibly toxic when bad ideas (like the fear of childhood vaccination) proliferate and are given unwarranted legitimacy simply by virtue of repetition. However, it can also work to the good, and the very "derivative" nature of much of the current steampunk genre has the potential to help make the imagined worlds described therein seem more plausible—less like fantasy and more like something that everyone agrees could be. Furthermore, bad writing is subjective and from the perspective of social change, accessibility is almost as important as content. A steampunk comic book rich in ample bosoms and simple sentences may not rise to the level of literature, but if it popularizes the ideas presented in books by "better" authors if the cost of that popular appeal (e.g., the eating disorders prompted by those bosoms) doesn't outweigh the progressive message, then it still must be counted a success.

People are enthusiastic about the genre so there is more of it being written—and the fact that much of it is derivative and forgettable doesn't make it any different than most of the rest of the stuff at the bookstore. We're still reading Jane Austen and Emile Zola while most of their contemporaries (and they had many contemporaries) are forgotten—it's unfair to expect anything different from the popular literature of today.

At this point it would be fair to point out that this isn't new ground. There isn't much that we've said about steampunk that couldn't just as easily be applied to its nearest ancestor, cyberpunk. But as William Gibson discussed in a recent interview in the *Paris Review*, there is a sense that cyberpunk was too easily appropriated by the very power structures it sought to undermine and that quickly it was transformed from something that urged us to "hack the system" into an aesthetic used to sell us more unfair mobile phone contracts on hardware manufactured under horrific conditions in third world factories. To be fair, we think that the story isn't out on the world that cyberpunk promised us and that the apparent dominance of information megacorporations like Facebook, Apple, and Google may yet be thwarted by the vision of DIY hardware, open-source code, and insurgent hackers. However, even if cyberpunk truly is dead, we nonetheless have reason to hope that steampunk won't share the same fate—and for reasons that force us to admit our own misjudgment.

We've been in the camp that sneered at the people who dressed up in steampunk costumes and assumed artificial personas in the style of a roleplaying game. We've wanted the "costumes" of steampunk to be fashion and for people to interact with reality as Jane Smith rather than Airship Captain Euphemia Mountebank. We still want that, and we still believe that steampunk will eventually wither if its practitioners consistently favor game-spaces created within convention hotels and populated with characters over messy social geographies populated by people. We got this wrong; and are increasingly convinced that some measure of apparently escapist fantastical roleplaying is actually protective against the malign influences of mass commercialism. We suspect that the apparent dissonance between the literary and role-play-heavy social cultures of steampunk is actually a strength rather than a weakness. Steampunks can be sold all sorts of little bits of crap, all kinds of movies and music, and they'll click "like" on pictures of kittens in goggles with reckless abandon. They don't, however, seem particularly interested in buying someone else's vision of what their fantasy personas ought to be. Efforts to sell particular identities (e.g., like those marketed at Steampunk Emporium and Clockwork Couture) may have sold many individual items but we've yet to meet a steampunk who's taken one of those identities wholesale. Similarly, while steampunks are happy to play genre-specific RPGs like *Unhallowed Metropolis* and *Space:1889*, they tend to keep their characters in these games separate from the personas they inhabit as "steampunks" per se. It seems to be a critical aspect of steampunks' characters that they inhabit their own unique world or a world created with just a few other members of, for example, an "airship crew." These worlds remain the inventions of their participants and even if Jane Smith bought her outfit at Hot Topic, there are no corporate sponsors of the world she's invented to inhabit as Airship Captain Euphemia Mountebank—and in that protected space there is at least the potential that she can imagine a world without shopping malls and a dominant automobile-petrochemical complex. More importantly, there is a better than average chance that she'll carry some of that dream back into her real life.

A related issue is the fact that many self-described steampunks are not just disinterested in steampunk as a vehicle for social reform, but are actively opposed to it. Any topic even tangentially related to real-world social or political issues (even historical ones) is guaranteed to provoke loud objections that "steampunk shouldn't be political!" In the defense of this escapist argument is the reality that the heterogeneity and size of the population interested in steampunk ensures there will be two (or more) intensely opposed points of view over even the most innocuous political positions (e.g., that Abraham Lincoln addressed a hideous moral stain in America or that women's manumission was a positive development) and that maintaining something like a functioning collegial community is challenging in the face of such dissent. Civility is easier to maintain when the most intense argument (carried out ad nauseum but apparently without ever boring many steampunks) is over whether a particular object/song/movie/book/commercial/etc. meets an entirely arbitrary definition of steampunk. Unfortunately, while the adherents to the willful escapist position do suggest a less contentious community structure, they also undermine the beliefs and actions of those who do want to use steampunk as a platform for social change. We'll admit that we don't have a great solution to this problem—ultimately we can only hope that the community is tolerant and self-policing enough to endure despite some people wanting to talk "steampunk politics" and some people not. However, in one sense, the escapists are completely irrelevant—namely, with respect to steampunk as a visionary movement. If an escapist wishes to shout down steampunk as apolitical but is willing to participate in a fantasy space in which European explorers interact on equal terms with women and indigenous peoples and in which pirates are ethically justified in robbing from exploitative industrialists—well, he can continue to believe that he isn't endorsing a political movement, but for all the reasons we've discussed above, he's still helping.

The preceding paragraphs seem to indicate that we're off the hook no matter what we do. Most steampunk writing can be terrible, the steampunk aesthetic can be applied to market endless quantities of cheap plastic crap, and steampunks themselves can be aggressively apolitical and despite all that we'll still help to shape a better tomorrow by imagining it. This is clearly false and the elephant in the room is the nature of the vision that steampunks embrace. The power of imagination does not absolve us of responsibility to be mindful of the past and to consider the nature of the future that we want to construct—it does just the opposite, and we'll admit that this causes us to temper our optimism. If the fantasy worlds of steampunks embrace toxic social constructions then those visions are guaranteed not to create a different better world but to replicate (or even worsen) our own. There are vocal elements within steampunk who genuinely look back to an imagined nineteenth century of militarism, imperialism, racism, and corrupt gilded age capitalism with misplaced fondness. We don't know whether this is a function of ignorance or sociopathy, but the fact remains that many steampunks are not particularly suspicious of their nostalgia and do long for the "good old days," albeit ones enhanced by shinier "stuff."

Ultimately it's up to all of us to determine the nature of the worlds envisioned by steampunks. It's within that landscape of imagination where the battle for steampunk's soul will be waged—is being waged; and where our question will ultimately be answered as to whether steampunk matters. However, we are confident that it at least *could* matter, because even when appropriated by corporations, steampunk has the unique potential to allow us to visualize worlds different from our own, but similar enough that we could make them a reality. Even if we believe the people who argue that steampunk is only an aesthetic, a style, a literary motif—we have to remember that aesthetics have power. There is a reason that tyrants and totalitarian regimes murder or subvert artists and writers. A novel can change the world, because it can reconstruct the spaces inside our heads. Steampunk allows us to imagine change and to build invisible cities that might be. If we look around our real world and don't like what we see, then we must start building someplace else, someplace better. What better place to start building that better tomorrow than in the landscapes of imagination? Where can real world change begin other than in the mind?

I run across Unwoman everywhere, and it pleases me every time. I've seen her performing at a steampunk convention in Portland, a coffeeshop in Canada, a punk club in Berkeley, and on the streets of Oakland during the historical November 2nd, 2011 general strike. And she looks just as calm and collected playing for civil disobedients as she does in a hotel ballroom. I spoke with the cellist and songwriter about science fiction, the steampunk scene, and what it means to be a hardworking artist living by her wits and the grace of her many fans. She can be found online at UNWOMAN.COM

UNWOMAN

An interview by Margaret Killjoy
Illustration by Sarah Dungan

***STEAMPUNK MAGAZINE:** Can you tell me a bit about how you got into steampunk, where you were coming from, what appeals to you about the culture?*

UNWOMAN: I've been into goth and industrial subcultures since my mid-teens, but never wanted to make music that fit into either of those. Steampunk welcomes the dark beauty of the goth esthetic, and DIY mindset and dystopian futurism of classic industrial culture, without having a concrete idea of how its music should sound. And since I write songs without time markers, play classical instruments, and produce electronic beats, it makes sense for me. And the steampunk community has really embraced me. I've always loved retrofuturistic styles and science fiction and wearing corsets, and I get to enjoy a lot of what I'm into at the steampunk conventions at which I perform.

***SPM:** What kind of science fiction are you into?*

UNWOMAN: I don't know exactly what kind it all is. My favorite current author is probably Charles Stross. Of classics I love Alfred Bester, Arthur C Clarke, Isaac Asimov, and some Philip K Dick (there is a lot of bad Dick, but good Dick is great.) Of things that can be called steampunk, I recently read and loved Paolo Bacigalupi's *The Windup Girl* and am enjoying Gail Carriger's Parasol Protectorate books, and of course I tell everyone I know about your book *What Lies Beneath the Clock Tower*, which is absolutely brilliant. I have a huge soft spot for dystopian futures—of course Margaret Atwood's *The Handmaid's Tale* is a favorite.

***SPM:** As a con-going performer, you've probably had a chance to see steampunk develop, or at least its regional variances. What can you say about those things?*

UNWOMAN: I remember a huge sense of excitement at the first Steamcon, in October 2009. It was one of the first major gatherings of steampunks specifically calling itself steampunk (you probably have better history on this), certainly the first I attended. It was really thrilling to be among so many well-dressed and friendly folks, really eager to get more involved. In the two years since I've gotten pretty used to these conventions and that initial thrill has worn off for me, but that's not a bad thing. Now I am honored to be able to calmly observe fresh faces with similar excitement for their first steampunk conventions and I feel completely at home performing at them.

I don't see a lot of regional variance, I tend to see similar modes of dress and convention panel topics everywhere I've been, though the farthest East I've played a specifically steampunk event is Oklahoma City (I did play Eccentrik Fest around Halloween 2009, put on by the Davenport Sisters of Clockwork Cabaret, in North Carolina, but it was too Halloween-y to gauge local steampunk style) so I haven't seen how they do it on the East Coast first hand (yet). One thing that's true of course is in scenes with less steampunk presence, styles are more varied and less exclusively steampunk—there are more burlesque, circus, goth, dieselpunk, and other retro styles represented. I like these too, and believe steampunk needs other influences to stay interesting.

SPM: *Your latest album is* Uncovered, *which is of course a cover album. What is it that appeals to you about covers? I hope it doesn't sound like a stretch, but I can see it relating to steampunk, too, the re-appropriation of culture.*

UNWOMAN: Learning covers was a really important part of my early musical development. Though I've been writing my own songs for nearly two decades now and have four original albums out, I've kept playing other people's songs. At this point, I choose songs to cover if one of these two things applies. First, if I see a way of interpreting a song loyally but from my own, completely different, perspective, or just with my own sound; second if a song makes me cry and I want to master the emotion. Examples of the second include Amanda Palmer's "Ampersand," Antony & The Johnsons' "Cripple and the Starfish," Front 242's "Crushed" (which appears on *Uncovered*), and most recently Peter Gabriel's cover of The Arcade Fire's "My Body Is a Cage," of which I made an unauthorized remix for bellydancers rather than covering it. I'm fascinated by music's power and am always looking to explore that power by learning the songs that touch me most.

I don't know if covering songs is particularly steampunk, but I think covers do well in a music genre that doesn't exist or isn't (yet?) defined by its sound, because the songs give people an anchor they recognize, allowing musicians to play with styles that don't fit a commercial mold.

SPM: *I'd love to hear your thoughts on making it as an entertainer right now. What are some tricks you've developed, things you've learned the hard way? What's involved in, for example, releasing an album that consists of other people's songs?*

UNWOMAN: I can answer the second question—releasing covers is very easy because unless you're syncing to visuals, you don't need permission, you only need to pay royalties. For this album I used a service called limelight (SONGCLEARANCE.COM) which does all the research and paying of rightsholders for me for a small fee per license—completely worth the time it saved.

The first question is so open ended! I definitely haven't made it, but I might be in the process of making it, so I can spout a little advice I've picked up: be enthusiastic—you have to love what you're doing 90% of the time or you're in the wrong line of work. Say yes a lot. Be courteous, professional, and punctual. Drop the balls you're juggling as rarely and as gracefully as possible. Be generous with your time on other people's creative projects without overcommitting yourself or being exploited. Say no a lot. Kiss a social life involving just "hanging out" goodbye but get comfortable with networking. Don't take it personally if friends don't support your music. Be okay with doing things you aren't good at, like self-promotion and booking shows for yourself. Hear every criticism without letting it push you away from your vision. Keep your high standards but don't let those standards prevent you from producing. Keep an audio recorder (or a recording app on your phone) and/or a notebook with you at all times and jot down every idea for later. Don't wait to be discovered. Let fans and friends help you. Put out a hat for tips any time it's not tacky to do so—you will actually sell more merch if you ask for tips. Make personal connections with fans and reply to their emails and tweets but maintain whatever boundaries you need without apologizing. Do everything you can yourself (in my case, all recording, production, web design, booking), and for everything you can't (photography, mastering, video work), pay awesome friends (in money or trade) to do it rather than buying services from faceless professionals who don't get your aesthetic. Don't play first-year conventions unless the people running them have great reputations as promoters or pay you in advance. If half of the strangers in the room you just played bought your CD at the end of the show, you're doing it right.

The Mechanic

Marie Morgan
Illustration by Sergei Tuterov

I'M TWO HOURS TOO EARLY. THE POCKET WATCH IN MY HAND goes *tick, tick, tick*, and I almost want to take it apart to make sure it isn't slow. Two hours isn't bad, really. My first year, I spent the better part of a month in this spot, gazing longingly upward and counting the ticks.

The shaft goes so far up that the orange flame of my lantern doesn't reach the top. It perfectly illuminates years of rust and grime on the metal walls, though. I have each stain memorized. If I squint, the one across from me looks like a chicken.

Tick, tick, tick.

I smooth out my dress for the thousandth time, noticing a grease stain on the hem. A scowl twists my face, but it's inevitable. Grease gets on everything down here. Trevion knows that; he shouldn't get mad at me for dirtying his present.

Then I can't hear the ticking of my watch anymore. It's drowned out by the sound of hundreds of tiny metal feet.

The Machine's horde of mechanical servants fills the corridor. Cat-sized metal spiders, they crawl over each other in a constant writhing mass. I assume they were sleek and black once upon a time; now they're rusted and leaking grease. Half of them are patched together with spare parts: pipes and antenna in place of legs, mismatched metals grafted onto their bodies. They limp and creep but stop just short of the light of my lantern.

"How nice," I say. "A sending-off party."

"YOU SEEM EAGER ENOUGH TO LEAVE."

"Of course I am."

I address the horde of spiders, and they shift in the darkness, clicking and clanking. It's easier to look at them than stare into nothingness. Three years here, and I've learned the tricks to talking with Mac's disembodied voice.

"PERHAPS I WILL NOT LET YOU GO THIS TIME."

I look into the darkness, into the bowels of the mechanical beast, my prison for the past three years.

"Oh, stop whining," I say. "It's only for a day."

The shafts and tunnels rumble with his irritation.

I ignore him and turn back to my pocket watch.

At the five-minute mark, I start to climb the ladder. Up and up it goes, until the point where I would break every bone in my body if I fell. That's how most Mechanics die if insanity doesn't get them first: plain old accidents. Sure, a few manage to pass away from old age, but people don't call it lucky. You're effectively dead from the moment you climb down here.

I see the thin slits of light around the seal and climb faster. Then I slow back down, because I'm used to climbing in trousers, not a dress. And I wasn't going to slip and die today. Tomorrow, sure, but today would just be disgustingly ironic.

By the time I reach the top, it's two minutes and counting. I clutch the rungs tightly and wait. The next two minutes feel longer than the past two hours, but eventually it happens. There's a dull, deep thud as the seal unlocks, and then the gears crank and lift it open.

A hand appears before me, and I take it. A moment later Trevion pulls me into his arms.

He whispers my name, and then we're kissing. By the time we break apart, I'm gasping for breath. And it's fresh air I suck in, not the stinking, steaming fumes from below.

He looks me up and down like he's trying to memorize every inch of me.

"You got grease on your dress," he accuses.

I roll my eyes. "There's grease on everything down there."

"No matter. I'll buy you a hundred dresses."

"Yeah?" I snort. "I'll look absolutely ravishing as I fix the boilers."

Then Trevion Cale, the city's chief engineer, puts his arm around me and leads me back to humankind.

The first thing we do is make love. That's been number one on our to-do list for every one of my holidays. Then it's dinner, and not the dry rations and water I survive on below. It's seriously fancy. There's a chandelier hanging from the ceiling that casts the opulent dining room in gold. The table itself is groaning under the weight of the feast upon it. There's no way Trevion and I could eat it alone. A dozen people couldn't have managed it.

I try a little of everything: lobster in butter, honeyed ham, truffle cheese, strawberries and cream. I savor each bite and try to store it in my memory for the coming year. On the opposite end of the long table, Trevion sips wine and gazes at me.

"You make me nervous when you do that," I say.

"I see you so little. I want to drink you in while I can."

I hold out my hands to my sides. "Drink away."

He smiles. I start a bowl of cream of mushroom soup.

"You're so beautiful," he says.

The soup nearly comes out my nose.

"I've loved you since the day I saw you," he goes on. "Back when you were a lowly acolyte, and I was your master."

Now I laugh. He stares at me in shock.

"You're so ridiculous, Trevion. That's what I love about you."

He recovers his smile. "I'm glad I amuse."

We go to see my family afterward. The house is not the cramped apartment I grew up in but a mansion in the nice part of town. They have mechanical servants now, the kind that shine and barely smell. I'm glad my position has given them something. I still remember the way they wept when the Machine called my name.

Wealth hasn't changed them. My little brothers and sisters run around wildly and shout their excitement. My mother hugs me and breaks out the beer. Trevion is an outsider now, bewildered by old family jokes and lower-class customs. My youngest brother climbs all over him as we talk, and my mother is so thrilled to see me she doesn't notice. I catch Trevion's eye and grin.

There's no way they can tell me all that's happened in the past year, but they try. I can't believe how big some of the little ones have gotten, or the neon shade of green my brother has dyed his hair. One of my sisters is married now, and I never got the chance to threaten her husband. So many things I've missed.

My mother asks me how I'm doing and then looks away in embarrassment, like she knows it's a terrible question. I tell her I've gotten used to it, which I guess is true enough. She has more wrinkles than I remember, and I'm just glad she doesn't have to worry about putting food on the table anymore. I hope my siblings are taking care of her. I'd like to think they are, but I feel like I barely know them anymore. They're all so different.

I guess I'm different, too.

When I climbed down the ladder for the first time three years ago, it was a miracle I didn't slip and break my neck. I was shaking and crying, my hands wet with tears and snot, and I could barely see where I was going. Plus, I was wearing a dress.

I reached the bottom and clutched my pack of belongings like it was my life. Then a metal spider scurried towards me, and I shrieked and scrambled back. It stretched out a corroded leg on which a lantern dangled.

"TAKE IT," the voice boomed

I flinched back with a squeak and snatched up the lantern. The little spider clicked its way down the hall.

"THIS WAY."

I followed obediently, trembling and casting nervous glances into the darkness. The place was a maze of corridors and tunnels, ladders and crawl spaces. For a brief moment I worried about getting lost, but so what if I did? I was trapped down here anyway.

The spider stopped in front of a metal door.

"THESE ARE YOUR QUARTERS."

I wondered if I should say thank you, but I was too scared to utter a sound. I reached for the door handle, wanting to lock myself away.

"NOT YET. I AM NOT FINISHED CLEANING."

I cringed. Everything it said sounded furious, like a proclamation from hell.

For a time we waited, and I became aware of a sound behind the door like rain hitting a tin roof. Then the door burst open, and dozens of metal spiders crawled out. They carried between them the bloated body of Oral Ornez, my predecessor.

I heaved and swallowed back bile. *Don't throw up*, I told myself. *You'll have to clean it.*

"THERE. NOW IT IS READY."

The sound of the spiders retreated as they carried their gruesome burden away. My stomach was still rolling, and I tried to banish the image of Oral's blotchy, lifeless face from my mind.

"YOU WILL TAKE YOUR NECESSARY SLEEP NOW," the Machine said. "TOMORROW MORNING YOU BEGIN WORK."

I entered the room of the dead man, curled up in the corner, and cried myself to sleep.

My first months on the job were all the same. The Machine told me what to do, and I did it. Whether it was fueling the boilers that powered it or repairing parts and mechanisms that broke over time, my life was no longer my own. From waking to sleeping, I went about my duties like a zombie. I was constantly sweaty and greasy, and I stopped bathing after the first month. What did it matter? There was nobody down here to look nice for.

Every day I thought about my family. I hoped my mother was doing alright, since I wasn't there to help her look after the little ones. When my work was mindless I would write letters to them in my head, even though I had nothing new to tell them. As the months passed and my memories grew fuzzy, I couldn't always picture their faces. Sometimes I cried. Okay, I cried a lot. I wondered if they missed me as much as I missed them.

Time passed slowly, and the life I'd once lived began to seem like a dream. I barely felt human, like I'd turned into some kind of mechanical automaton like the spiders that served the Machine. I was so damn lonely. I grew deliberately careless atop tall ladders and under heavy machines. My first holiday seemed too far away to bear.

After half a year, I realized what was happening. Everyone remembers the Mechanics who come out on holiday hollow shells of their former selves. The Machine passes its madness to its Mechanics, they say. I say anyone's bound to go mad being trapped down here for 364 days out of each year. But I didn't want that. I didn't want to come out on holiday and have my mother barely recognize me, have her weep and say I would've been better off dead.

I took a bath. I wasn't going to do it daily—that was pointless down here—but once every week or two would be good enough. As the Machine barked its orders to me, I started thinking about what I could do to stay sane. It wasn't like staying busy would help; I got enough of that already. Although it would be nice to think I was making a difference instead of being one temporary solution in a long line of many.

I was replacing the latest section of pipe that had rotted through when the spider that was carrying my toolbox slipped and sent them crashing across the floor. Not that I blamed it. If I was missing half my legs, I'd be a little clumsy too.

"Hey," I said. "Do you want me to fix those?"

The spider scrambled to recollect the tools.

"FIX WHAT?" the Machine boomed.

"Your little spider things. They're all messed up. Do you want me to fix them?"

"THEY ARE NOT VITAL TO THE RUNNING OF THE CITY."

"Sure, but do you want them fixed?"

It was silent for a while, and when it spoke, somehow its voice didn't seem so loud. "WANT?"

I finished attaching the pipe and hopped down. "Like how you want all this other stuff fixed."

"THOSE COMPONENTS MUST BE FIXED. OTHERWISE I WILL DIE AND THE CITY WITH ME."

"Yeah, but... you must want things. Like..." I paused. If the Machine was sentient enough to go mad, it must be sentient enough to have wants. I tried to think of an example.

"Like how you only want one Mechanic and don't let anyone else in."

"THAT WAS NOT MY DECISION."

This time its voice was so loud the entire room shook. I clung to the pipes to keep upright as dirt and roaches shook loose and fell from above.

"YOU PEOPLE MADE IT THIS WAY. YOU WANTED ME DEAD. I CAN ONLY ALLOW ONE. I WOULD NOT BE ABLE TO KEEP MORE FROM KILLING ME. I MUST KEEP MYSELF AND THE CITY ALIVE."

The pipes swayed dangerously and groaned. Bolts popped out and shot past me. The entire room felt like it was coming apart.

"Alright!" I shouted. "I get it! I'm sorry!"

The rumbling stopped and the room settled. I took a deep breath and straightened up.

I should have been cowed. I'd just been chastised by the voice of the devil. But it had sounded more like a temper tantrum.

"So do you want them fixed or not?"

There was a long pause.

"YES."

The clocks move cruelly fast on holiday, and soon it's time for public duties. You'd think I fulfill my duties the other 364 days of the year and could spend today with Trevion and my family. But no, even on holiday my time isn't completely my own. I shake hands with Important City Officials, and they pretend not to notice the grime permanently stuck beneath my fingernails. One mentions how I must like being up here better than with the riffraff below. I don't know if he's being insulting or just stupid, and I'm taken off guard and too angry to speak. Trevion dismisses him coldly, saving me the trouble. Not that Trevion's ever been poor, but he loves riffraff like me anyway. Then he, my family, half a dozen damned officials, and I take our seats on the balcony of city hall and watch the parade.

It's still strange being up here. I spent my life watching parades for previous Mechanics. Now I watch ones in my honor.

The floats are bright and colorful, shaped like animals, castles, islands, and everything imaginable. Some roll on wheels, others walk on legs. Steam puffs out exhaust pipes behind them. A one-man orchestra actually pedals his contraption, pulling levers and pushing buttons to operate the musical instruments. Another float has a completely see-through chassis, and I watch the colored gears turn within. Still another is shaped like a metal dragon and belches fire. It would remind me of the machines underground if it wasn't so clean.

"I designed this last one," Trevion says.

I lean forward to see. There's a large gap between his float and the one before, so that the whole crowd turns in anticipation.

It's shaped like a flower. I don't know what material the petals are, but they look like pale pink glass. They unfold before the crowd with the turning of gears, revealing a gold metal center that shines in the sun.

Then it shoots bolts of electricity into the air.

The crowd cheers, and I laugh with pleasure. It was expensive, elegant, and completely over the top—Trevion to a tee.

He takes my hand, and we share a smile. I wonder how long he spent building the thing. It almost makes having to sit through the parade worthwhile.

Mac hates parades. He says they give him headaches.

"What is it?" Trevion asks.

I shake my head. "Nothing."

By my second year, I was talking to the Machine all the time. There was no one else to talk to, and it helped me feel less lonely.

"So we're running from the coppers, and there's this fence in the way so we climb it, only Po's pants get caught at the top." I held out my hand, and a spider put a set of pliers into it. "And he can't get it unstuck! So he's got to kick off his boots and slide out of them." I clipped the wire and handed the pliers back. "So here we are running down the street, him in his drawers! It's lucky we found a hiding spot or he would've been in more than one kind of trouble."

I reattached the wire and climbed down the ladder, wiping the sweat from my brow.

"I DO NOT SEE THE HUMOR IN THIS."

"No? Well, I guess you had to have been there." I took a swig from my water bottle. "So what's next?"

"SECTION 42. THE PUNCHCARDS NEED RE-LOADING."

I followed the clicking spiders, which were doing much better since I fixed them up with parts from the scrap pile. They weren't works of art, but if anyone was good at making machines from junk, it was me. I didn't think any of the other Mechanics could've managed it, and I was sort of proud.

I wished I could tell my mother about it, or show my siblings. They would get a kick out of the creepy things. I would have to remember to mention it next holiday.

But I wasn't going to let that depress me. A song my mother used to sing popped into my head, and I started humming the tune, throwing in the words as I remembered them.

"WHAT IS THAT?"

"What?"

"THAT NOISE YOU ARE MAKING. HAVE YOU GROWN DEFECTIVE?"

"What? No. It's just a song." I frowned. "Haven't you ever heard music before?"

He was silent for a while. "NOT FOR A LONG TIME."

He'd gotten me feeling sorry for him again, the big lug.

"Hey, so I've been thinking. You need a name."

"I AM THE CITY. WE ARE ONE IN THE SAME."

"Yeah, that's great," I said. "But I'm thinking 'Mac.' Short for Machine. What do you think?"

"IT SOUNDS RIDICULOUS."

"Oh, come on. Give it a chance."

The nearest boiler shot out steam with a "pfft" of dismissal.

"Fine. Be that way."

I reached section 42, and the spiders led me towards the engine in question. I was grappling with the long sheet of paper for the punch cards when Mac spoke again.

"I WANT YOU TO MAKE THAT MUSIC NOISE AGAIN."

I tried not to smile. "What do you say?"

The room rumbled with his annoyance.

"PLEASE."

My holiday is over all too soon, and I'm bidding my family a heart-wrenching goodbye. My siblings make brash promises of all the things they'll have accomplished by next year, not knowing how much it hurts me to think I won't be there to watch them. My mother just tells me how much she'll miss me, and I can tell she's trying desperately not to cry. I hate this part more than any other time in the year. It's almost enough to make me understand why some Mechanics didn't come out on holidays. Almost.

They give me a parting basket of food and beer, and then Trevion and I start back. We walk the long way and cling to each other for every second of it. I want to memorize the feeling of his arms around me. It's going to be a long year before I get any human contact again.

We still have a long way left when he stops.

"We need to talk," he says, and his words are clipped and serious. I'd been expecting tender, flowery declarations, and I frown.

"Okay..."

"I have something for you."

He reaches into his bag. Another present? As much as I appreciate the gesture, he always gives me the most impractical things. But that wouldn't explain his business-like tone...

He pulls his hand out of the bag, and I stare at what he's holding for a long moment, willing it to make sense.

"Is that a gun?"

It's sleek and bronze, unlike any gun I've ever seen. Decorative designs are engraved into it, looking almost like lace, and I can't find any place to put the bullets.

"Yes, and it's priceless, so be very careful."

He puts it into my hands, and it's much heavier than it looks.

"I hate to break it to you, Trevion, but a gun's an even more useless present than a dress. Just who do you think I'm going to need to shoot?"

"The Machine."

I open my mouth, but no words come out. The realization seems to leave my brain and crawl down my throat, choking me.

Trevion smiles, thinking me in awe. "Brilliant, isn't it? The city's brightest minds have spent the past decade working on it in secret. With this, you can destroy the Machine's main analytical engine, effectively killing it."

"But that will shut down the entire city!" My voice comes out far too loud, and I have to force myself to sound calm and reasonable. "I mean, everyone knows that. He's the only thing keeping the whole place running, right?"

Trevion gives me an odd, worrying look. "He?"

"The Machine."

"It's just a machine. The engineers and I are confident we can replace it—and compensate for its absence in the meantime."

I fiddle with the sleeve of my dress. "That still sounds awful risky..."

"It's what needs to be done. We've lived under the Machine's tyranny long enough, don't you think?"

"Well, I don't know if 'tyranny' is really the right–"

"It's mad." He puts his hands on my shoulders and looks at me intently. "You know that better than anyone."

I look away from him. Inevitably my gaze falls on the city streets, with the houses full of happy families asleep beneath a sky of twinkling stars. To be up here all the time...

"What if you're wrong?" I ask. "Let me see the plans. If you can really–"

"We've been working on this since long before you became the Mechanic. Don't you trust me?"

"Of course. I just..." I struggle for words. "It's not the Machine's fault he went mad."

I can tell this isn't the reaction he was expecting from me. His brow wrinkles, and I feel awful for disappointing him—but not as awful as I feel about the thought of killing Mac.

"I don't think it was ever meant to last this long," Trevion says gently. "It's broken, plain and simple. You've seen how paranoid and miserable it is. Killing it now would be a mercy."

But Mac doesn't want to die.

Trevion takes my hands in his, clasping our fingers around the cool metal of the gun. "Think of it. No Mechanic will ever be drafted again. No one else will have to be separated from their family, worked until they're dead or worse." He kisses my forehead. "You'll be free. And you and I can be together. Always."

I can't meet his eyes, but it's his touch that makes up my mind, his warm skin and strong embrace. I need the touch of another human being more than a single day a year.

"Okay..." I say. "Okay."

He puts his arm back around me, and we finish our walk.

I kiss him goodbye and climb down the shaft, back into the hot, stuffy air, the smell of grease and metal, and the ever-present darkness. It's all so familiar, almost homey in a way, but the farther I go the more oppressive it feels. Years of isolation weigh down on me. But now I can go back to the light and the people I love. All I have to do is pull a trigger.

"I AM GLAD YOU ARE BACK," Mac says.

I perk up. "Really?"

"YES. A GASKET IS LOOSE IN SECTION 14. YOUR IMMEDIATE ATTENTION IS REQUIRED."

"Oh."

The gun is in my bag, and I put it in my quarters along with everything else I've brought from above. There's no need to do anything right this second.

I see to the gasket and then say goodnight. Mac knows my "BEHAVIOR IS MORE ABNORMAL THAN USUAL" after holidays, so he doesn't bother me. I've been awake for the past twenty-four hours, and sleep sounds wonderful, but I don't think I'm going to get any. Not with this on my mind.

I lie down on my cot and stare at the wall where some prior Mechanic gone mad scribbled nonsense. I'd always dreamed about fixing Mac—really fixing him, not just these temporary repairs that keep him going. I used to think I could convince him to let other people work on him. He'd do better with more people to talk to. Being alone down here isn't any better for him than it is for me.

I wonder if Trevion will still love me when I'm no longer a tragic martyr, when he'll have to put up with me day after day. We've been together two years and have seen each other for a grand total of three days. Does that make our relationship stronger or weaker?

Not that he's really the one asking me to do this. It's the city officials who had the gun made, the stupid rich bastards who control everything but Mac. They say it's the Machine's tyranny we live under, but really it's theirs. At least, that's what I think.

I think of the Mechanics before and after me. Well, not after me, not anymore. I'll be the last. No one will ever be ripped from their loved ones and forced to live down here again. I guess that means I'll make a difference after all.

Then I think of my mother. I'll do it for her, if for no other reason.

But not tonight.

Morning comes, but I'm not ready. I need more time to work up the courage. Trevion calls Mac "it," but I still think this is murder.

The next day. I get as far as picking up the bag with the gun. I'm still not ready, but tomorrow I have to be.

I'm too busy today, and too exhausted. Tomorrow.

Just one more day. Mac's grouchy. I want him to be happy in the end.

I can't keep putting this off.

One more day.

Fuck it.

I grab the bag and stride down the corridor. Spiders scatter from my path, and I seize the first lantern I come upon. I can feel the weight of the gun, and I curse it and every engineer who helped build it. Then I curse myself. I'm so selfish. And a traitor. This is the wrong choice, I'm sure of it. But I'd rather live in regret than this endless indecision.

Tears are streaming down my face, but I'm too angry to care. I'm ending this now.

"WHAT ARE YOU DOING?"

There's a hole at the edge of the Machine's insides. Even Mac doesn't know how far down it goes. I pull out the gun—and toss it down.

I never hear the sound of it hitting the bottom.

"WHAT WAS THAT?"

I let out a deep sigh of relief and clutch the wall. My knees are weak.

"Nothing. Just throwing out some junk."

"TRULY?"

"Really. It was junk. I'm going back to bed."

I stand there for a while instead, my mind racing and my heart still pounding.

"Tomorrow we need to talk," I tell him. "There's some things I want to do… differently from now on."

"DIFFERENT?"

"Yeah."

He thinks for a while. "I WANT THINGS TO BE DIFFERENT." A pause. "PLEASE."

"Good."

I start back to my quarters, and a deep moaning noise rises up from within the walls. I clench my eyes shut, because he's figured out the truth, and I should've just told him. Then the pitch of the sound changes, and I realize what he's really doing.

He's humming.

It sounds awful, but I smile anyway. The Machine passes its madness to its Mechanics, they say. Maybe they're right. I have no idea what I'm going to do next, but I'll figure it out.

I've got time.

Pearls Before a Sandhog

The Catastrophone Orchestra
Illustration by Larry Nadolsky

October, 189?, New York City

Mathilda was studiously ignoring Neal Lisst while watching Professor Calamity, who was engaged in a battle to stay upright that he was rapidly losing. She was too angry with Neal to speak with him, and the Professor was well beyond the point where he could engage in conversation.

She was mind-achingly bored. The court officers were discussing the merits of Dr. Mahoney's Mustaches Wax, so there was nothing for Mathilda to do but to examine the others. They reminded her of the frescoes of purgatory that adorned the Orthodox churches of her homeland. The couple dozen lost souls of night court were huddled in groups of threes or fours, spread out across the uncomfortable maple pews. The sickly gaslight and the lateness of the hour had drained all vitality from their faces. They stared at the judge's chair, currently devoid of its magistrate, their expressions vacillating between worry and boredom. She focused her gaze on a couple of Bowery Boys carving obscenities into the back of a bench with a stylish, pearl-handled pig-sticker. She was appalled by their poor spelling.

"It wasn't my fault," Neal said, looking straight ahead, transfixed like most of the others by the empty leather chair that presided over the room.

"Damnitall, Mathilda," Neal screamed when she stuck him with the hat pin that secured her mourning veil. Mathilda always wore a mourning veil and matching clothes but no one was ever sure who, precisely, had died, and most were afraid to ask the unstable vocalist of the Catastrophone Orchestra and infamous anarchist.

"I cannot abide wanton carelessness," she said. "Dock sneaks have a well-established rhythm, like the catastrophone."

"You know Pip. He's thick about them machines," Neal said, rubbing his white mohawk, "We'd got the loot we'd come for, clean as spit. The docks were clear and he could of pounced right out the hole in the fence. But you know how he is all cranked up. He wanted to see that new contraption all the papers were about, the 'badger,' so I got out with the scag, waitin' for him by the fence. It ain't my smear that he fluxed it all to hell. I got the valves and other gizmos, so why you sticking me?"

"I'll stick him too, as soon as he gets out of the tombs."

That was John Haviland's biggest architectural project: the New York Halls of Justice and House of Detention. Everyone south of Houston Street simply called it "the tombs." Built in the Egyptian mausoleum style, a short-lived fad at the time, the massive structure was as grim as it was poorly-designed. The holding cells were all underground—and since they were below sea-level, whenever it rained the cells would fill with a grisly and aromatic stew of storm run-off and human sewage.

Outside, it was raining.

While Neal and Mathilda were whispering back and forth across the disgraced alienist, Professor Calamity surrendered in his battle to remain upright. He slid from the graffiti-scarred bench like pudding, wet and formless.

"Come my love, you must rise," Mathilda said, grasping at Calamity's damp collar, the cardboard so saturated in sweat it simply disintegrated in her painted claws.

To make matters worse, the Professor, or more specifically, his stomach, started to groan, the noise somehow both eerie and repulsive. The two old Italian women sitting in front of the anarchist artists were so disturbed by the gurgling coming from underneath their bench they crossed themselves and moved to another row to await their relative.

The cocky thugs in their court officer uniforms broke off their conversation about facial hair grooming supplies and started to march down the narrow aisle towards the increasingly loud gastronomic rumblings.

Neal slid his hand in his left pocket where he kept his knuckle dusters and smiled. *Maybe this evening won't be a total wash,* he thought to himself as he watched the two officers walk his way.

But just as he'd decided on breaking the nose of the bigger of the two wraggles, and had begun to look forward to the impending brawl, Mathilda hissed instructions at him.

"Get the doctor out of here," she said.

Despite the widow's bustle, she wore her dress to accentuate the curves that had made Calamity originally fall in love with her at Bellevue. She was what, in another dress, would be called striking but in this one disconcerting.

Mathilda was queer, even for Manhattan night court. She went up to the man who had been going to have his nose shattered by Neal and set her gloved hand lightly on his truncheon. The other guard stared.

"I am so dry, I think I shall faint," she said, her whisper like smoke in the officer's ear, her body lightly touching his as her laced fingers gripped his leather club.

While the man uneasily gave her directions to the one working water fountain, Neal pulled the lanky Professor out from under the bench and pushed him through the swinging oak doors out into the hall, where Mathilda joined them seconds later.

The Tweed administration had defunded the maintenance of the building, since so many Tammany officials had spent their early years in the despised tombs, and the only gas jets that were lit were in the arraignment court. Even in the gloom, Calamity looked like week-old mutton: shiny and green. Neal was looking over his head, hoping the court jackals would be exiting. They didn't.

Mathilda took her lover in her arms and cooed in his ear.

"Get him what he needs," she said, dropping a well-worn syringe kit into Neal's coat pocket. "He needs to be sharp if we're going to get Pip home. It's late, so you'll need to go to Dragon Lilly's."

"Lilly? That's an hour or more and its pissing buckets out there," Neal protested, still angry he couldn't fight the officers.

"Pip isn't going anywhere for awhile. Now go."

Neal was headed for the glass doors when Mathilda called out after him: "What by hell is a badger, anyway?"

Jarek carefully packed his textbooks in the canvas sling his grandmother had made for him back before her coughing had forced her to first to Ward's Island and from there to Potter's Field. He crouched in the interior courtyard, his mouth watering, the smells of coffee and toast wafting out from the window of the noisy communal kitchen. He could hear "uncle" Brorys' guttural groans from the outhouse. Jarek knew he'd be late if he waited his turn, so he went to the corner of the courtyard and pissed on a dying mulberry tree. When Jarek returned, Anka was pawing over one of his textbooks, his favorite one. He ran over, his trousers still unbuttoned, and cuffed her hard in the back of her blond head.

"Don't touch those with your filthy hands," Jarek yelled, snatching back his precious book.

"It's not in English or Polish," Anka said, standing. "It's a made up language."

"It's algebra," Jarek said, checking his book for damage. "It's math. The letters represent numbers."

"Then why don't they just use numbers for numbers? That's chicken thinking."

Anka's older brother for her called from the window and she went back into the crowded tenement. Jarek always liked Anka and he knew she liked him. The two used to sometimes kiss in the basement of St. Stanislas on

Sundays, while they waited for their parents to finish with mass, but that was before she'd started work at the needle factory. Now she worked Sundays, and every other day besides, grinding fish steel on wheels. She came back late every night and Jarek needed to be up and out to the Settlement House by seven if he were to get breakfast. He had not seen Anka for over a week and it had cheered him to see her that morning, even if she had been pawing through his books.

The only thing Jarek liked more than Anka was his books. They were a promise and a challenge bound in strong cardboard and horse glue. The black and white of them were pure and he couldn't stand for them to be stained by the filth of the sandhog tenement. Everything on Avenue A was polluted by mud, coal dust, mold, or just plain old filth. Everything was an indistinguishable soup of gray-brown grime. It was a daily struggle to keep the books immune from the corruption of the Lower East Side.

After his mother had been struck dead by a drunken omnibus driver, it was this desire for incorruptibility that had turned him to religion. He still dreamt of that day, of his father kneeling in a puddle of horse piss holding the bloodied and muddied body of his mother and crying, endlessly crying while the traffic moved around him.

It took the families in the tenement a few days to raise the money they needed to bury her in the churchyard, days during which Jarek's mother was kept in the basement on a pallet surrounded by melting slivers of donated ice. The morning of the funeral, five-year-old Jarek crept down the coal-covered steps to the cellar to see his mother. His father and relatives were already at the church preparing—Anka's family was to bring Jarek along later.

He froze at the bottom of the stairs. Dim sunlight slipped through the coal chute and coal dust hung suspended like smoke in the light that fell on his mother. A rat perched on her chest, rubbing its mucky paws. He threw a chunk of coal at the vermin, and it scuttled away into the endless shadows of the low cellar. His mother's white dress, her wedding dress he was told later, was covered in the blotchy footprints of rodents.

Math was immune. It could not be corrupted by coal or vermin. It was pure in Jarek's eleven-year-old mind. The equations existed, removed from everything else.

"Hurry, my boy. You must have your oats," Willard Emmett-Hodges said ushering Jarek into the Settlement's main hall. "It's important—a healthy body makes a healthy mind."

The room was full of neighborhood children eating steaming bowls of oatmeal in silence. Influenced by the Quakers in Philadelphia, Mr. Emmett-Hodges encouraged the reformers of the Addams Settlement house to offer a free breakfast for the local children, Monday through Friday, so that the pure and nutritious oatmeal might help them improve their lot. The Settlement taught evening classes on citizenship, English, and the latest advances in health and sanitation to the mostly illiterate denizens of the Lower East Side. Mr. Emmett-Hodges had even started a project to take the most promising children of the crowded tenements and send them to good schools. Jarek had been the first one selected for that project and thus Mr. Emmett-Hodges took a keen interest in the boy's development.

"Did you do your studies?" Mr. Emmett-Hodges asked, pushing a bowl of oatmeal over to Jarek.

"I did," Jared said, wishing they served coffee, "but I was not able to finish my Latin translations."

"In God's name, why not? You know how important this is. It is not just your development but the whole model... You represent the model, Jarek. I have told you this before, my boy."

"I worked until my candle went out. I didn't have another. Latin is not until Tuesday so I hope to–"

"Martha! Martha, come here at once. It will be alright, Jarek," the man said, his smile returning.

Martha appeared in her starched uniform that made her look like a cross between a nun and a nurse, a large kettle of steaming porridge under her muscular arm.

"Can you please go to the pantry and see if you can find a gross of candles for our young Homer here?"

The kettle dropped to the table with a thud and Martha disappeared down a hall to fetch the candles.

"Would you like more?" the director asked, holding the ladle.

"No, thank you."

"Good. It is best not to over-tax the digestive system," the director said. "We are all very proud of you, Jarek. You see that, yes? You represent a chance for others. Maybe not these young angels here, but future children from the tenements. There is a struggle going on. Oh yes, an important discourse, my boy. The past is fighting tooth and bloody nail but there are other forces. Enlightened people who know that the future of this republic is a shining city on the hill calling to all. Do you know that when I started this settlement my very own parents threatened me? They did not do this out of malice, Jarek. They loved me like your mother loves you. But they wanted more for me. I had studied at Yale. Do you know where that is? Of course not. It doesn't matter. It was there, in those great marble halls, that it came to me. Education was the key. The golden key to end this degrading poverty and injustice. I have spent the last thirty years trying to see this mission through, my boy. To prove that education is the knife that will sever this Gordian knot of want. And you, you are the culmination of my work. Here are your nickels for the omnibus. How many nickels in one hundred dollars?"

"Two thousand," Jarek said, taking the fares for his ride to and from school.

"Oh my goodness! I almost forgot your pin."

The director reached into his silk vest and removed a silver lapel pin. It was shaped like a lighthouse and carried the motto of the Addams Settlement House, *Ad Lucem*. Jarek knew enough Latin to decipher the motto as "towards the light." He also knew that the director always took the lapel pin back at the end of each day because he was afraid someone from the neighborhood would rob the boy. Jarek believed the man was probably right.

It was a joy, taking the omnibus uptown to the Bentham Academy for Boys. It was slow, and often it was so crowded he had to ride on the outside, standing on his tippy-toes to clutch the window frame for balance. Some days he would arrive at school with his arms aflame with pain from the exertion. Even when it was freezing or raining, the freedom of movement never ceased to exhilarate him. For the first ten years of his life, he'd never ventured further than to the St. Mark's Market, a few blocks from his home. But when Mr. Emmett-Hodges got him into Bentham, he traveled the entire length of Manhattan. He passed through strange worlds and glimpsed other lives. He swept across Houston Street, seeing the crowds of Hasidim in their beaver hats and long beards selling pickles from barrels on carts while Sicilian crones sat silently behind racks of strange herbs that hung like dried miniature people. The omnibus crept along the oasis of Washington Square, where servants from the nearby mansions crossed under the arcade and into the green in their pristine uniforms. He floated along Broadway just as the vaudevillians still in stage clothes and smudged makeup made for the redbird elevated. He knew someday he would save up a couple of nickels to take Anka on this wondrous trip with him.

It hadn't been easy at first for Jarek to fit in at Bentham Academy. The other pupils had come from families of means and spoke a language different than his own—they shared the same words, but the meanings were different. The Settlement had given him clothes and hard shoes that pinched his feet, but a disguise alone did not grant him social access. It was three long weeks of eating alone before the first boy, Damian, came and sat with him. He had been new to the school too, having come from somewhere called London. Damian's queer accent and habits had also made him an outcast. Jarek and Damian, not quite friends, learned to share their loneliness together.

It was Damian who had helped Jarek catch up to the other students, but soon it was Jarek who was helping Damian and other boys, especially with their math. Jarek's math instructor, Dr. Pembroke, quickly recognized the young student's exceptional gift and lent his student biographies of famous scientists. Jarek especially loved Galileo, a fellow Pole.

Jarek took to his studies like a starving man to a holiday feast. He sampled everything without satiation. And luckily, it was not only at the Academy and the Settlement house where Jarek's gifts were appreciated, but also at home.

Jarek's father was not an educated man. He was a respected laborer in the tunnels, a sandhog, one who could fill a railcart in half the time it took any Pole or Swede who'd ever worked under the streets of New York. It filled Jarek with pride when he accompanied his father to the pub and the other men bought their drinks—he didn't even mind when the men made fun of his puny size. Jarek's father was a giant of man, bald-headed, with tree-hard muscles and a set of long, lovingly groomed mustaches. Jarek, on the other hand, was almost feminine and diminutive in size.

But while his stature and mustaches won him a certain admiration, what made Jarek's father a local hero were his actions on the notorious September 13th. In the neighborhood it was called Unlucky 13, and there was a special church service every year to commemorate the day. The sandhogs had been working, tunneling under the East River to set the pillars for the great Brooklyn Bridge. The pressure had built up to sensational levels.

The sandhogs had been an hour away from the lunch siren when a terrible popping sound echoed through the chambers, and one by one workers were forced out into the river by invisible power of pressure, slamming them and their tools through a hundred feet of river silt. The papers reported the astonishment of the picknickers at Battery Park who'd seen a half dozen men suddenly appear like a whale spouting out of the river. Most had been torn apart by the force, bloody body parts crashing back in to the slow, brackish water. Inside the tunnel, others had collapsed, blood gushing from their ears, the result of massive hemorrhaging in their brains. Jarek's father, ignorant of history and the alphabet, had known intuitively that the air needed to be released and so he'd opened the flush drain, straining at the giant wheel. The slush dam opened, flooding the tunnel. The sandhogs had been pushed like bobbers across the great river of water deep into the tunnel. Two had drowned, but the other thirty-five men had escaped, bruised and battered but alive.

Jarek's father never liked to talk about it—he'd lost a cousin and a few friends. But he did take the drinks.

Since the sandhogs were illiterate, it was Jarek who read aloud the pamphlets that earnest union organizers passed out from the shadows. With Pinkerton thugs and the police at their backs, the organizers relied on clandestine printing presses hidden throughout the cellars of the Lower East Side to spread their word. The workingmen's bars were the only safe place to read these announcements to the sandhogs. When there weren't pamphlets, Jarek would read two-penny newspapers to the drinking men. Lech, the crippled bartender, would lift Jarek onto the scuffed zinc bar so the boy's thin voice could be heard in the back of the underground tavern.

After they left the bar, Jarek would lead his giant father home and put him to bed. Jarek then would light his candle and translate Galen's histories or study the Periodic Chart of Elements until fatigue dragged him down into slumber. His father would gently wake him every morning at five with a kiss on his forehead. He would give him an apple or sometimes a chunk of cheese wrapped in a kerchief for lunch at school.

Jarek was munching on such a lunch while talking with Damian, who was fretting over the outcome of a recent algebra exam, when Dr. Pembroke crossed the leaf-strewn yard. The instructor was over-dressed for the unseasonably warm autumnal weather in his tweed coat, and a trickle of sweat inched its way through the folds of his ruddy neck.

"I need to talk to you at once," the mathematician said, excitement causing his voice to tremble. "I sent your recent exam to our Board of Directors. It was perfect. Beautiful, actually. Your understanding of quadratics is astonishing. Well, Mrs. Kensignton mentioned your fine work to a friend of hers... Mr. Henry Clay Frick."

Jarek stared at his teacher, not knowing what exactly the man was talking about.

"Frick, boy! He is the Carnegies' man. He is coming to Bentham to give you an award and a full scholarship. He has also insinuated he may give a substantial endowment to our school. He will be here tomorrow. The whole school will meet in the auditorium at 2 in the afternoon. Your guardian from the charity will also be here. It is a great thing, a very great thing indeed!" he said, clapping his hands together.

"Damian, your work needs... well, work. You passed but barely. Can't you be a bit more like your friend here?" he said, patting Jarek on the head.

"We need to tell someone your good news," Jarek's father said, picking him up and swinging him like a rag doll, a massive smile causing the tips of his mustache to dance. "Where are your books, son? Do you have any that have poetry?"

"I have a volume of Byron," Jarek said.

"Grab it and meet me downstairs in a few flicks," Jarek's pa said before disappearing behind the curtain that separated their room in two.

"They say you're some sort of genius," Anka said, sitting on the stoop next to her friend.

"Not really. How would you know anyway?"

"Nothing stays secret here, especially bad and good news. And that queer duck at the Settlement is telling everyone who'll listen. He is just honking on and on about it like it was the second coming,"Anka said. "You going to McGurks?"

"I guess my papa wants us to tell some people, but it already seems everyone knows. It's not that big of a deal. It was just a test and some rich suiter wants to give the school some money. That's all. I don't care about any of that. Not really."

"You should, Jarek. It's important. Not much good happens–"

Anka was interrupted when Jarek's father barreled out of the tenement in his church sleeves and, despite the heat, wearing a Moshe derby on his bald head. He looked like one of the boxers framed at McGurks, Jarek thought.

"Go ahead and tell her, don't be shy," Jarek's father said, placing a few scrawny peonies beside the humble gravestone. Jarek had only been to his mother's grave a few times. His father had always said it wasn't good to dwell on the past. Jarek felt awkward looking at the old veiled Italian woman who shared the graveyard.

"If you won't, than I will. Gracja, my dear sweetest, your boy had done it up good this time, alrighty. He made a name for himself at that mucky-muck school. He got those smarts from you, I know. You always said he was special and you were right about that. Not too shabby for a son of a stupid sandhog, eh? Just wanted you to know, that's all. Your boy brought you a poem. I remember how you always liked those type of words, never had the tongue for them but he sure does. Go ahead read some of that Brian for your mother."

Jarek pulled out his book as his father sat by the grave uncorking a bottle of beer. They spent an hour in the graveyard, Jarek reading poems under the leafless trees. Jarek's father took the flowers and put them in the beer bottle and then gently kissed the stone.

"Now we got to get you something for your big day, champ."

It was after dark by the time they returned from the tailors. Jarek's father had purchased a beautiful, cream-colored corduroy jacket for him to wear to the ceremony. Jarek loved it, but felt uncomfortable with the amount his father had paid. His father and the lazy-eyed tailor reassured him it was a real bargain, but Jarek didn't believe it.

"You really didn't need to do this, the omnibus would've been fine," Jarek said, looking out the window of the hack as it flew down the street.

"I wasn't going to trust the omnibus on this day, Jarek," Mr. Emmett-Howell said, securing a the Settlement's lapel pin on Jarek's white corduroy. "I have been waiting for this for a good time. The chance to show that, if given a half a break, the underclass can take their place with the rest of us. That is the promise of democracy. Mr. Frick being there… today is validation. Here we are. Don't be nervous, just focus on your studies and keep your clothes presentable."

The director took out a comb and ran it roughly through Jarek's hair. Jarek climbed the freshly-washed stairs to the school.

"My god! What happened to you Anka?" Jarek asked, looking at his friend through the wrought-iron gate that protected the yard during lunch break.

Anka's left eye was swollen and purple. Her face was covered in sweat from her hurried flight to the school.

"It's nothing. Ol' man Kreutz tried to keep me from leaving work. He's lucky he only got a shin kick, that stinkin' prick. Nevermind him. You have to come, Jarek. It's bad. Really bad," she said, holding back tears from her one uninjured eye.

"What kind of accident?" Jarek asked, feeling the judgemental eyes of the other boys upon him.

"It's some sort of new-fangled machine they've got in the tunnel. A badger or something. It's gone haywire. Bronislav is dead. Gregori's father. There was a collapse. Everyone is down at the site now. The bulls tried to keep them back but everyone is there. You need to come."

"What about papa?"

"Trapped like the rest of them. Thirty, maybe forty of them are stuck. Radaj sent me for you. He said you can read and that you're small enough. Come on. He gave me money for the way back," she said, showing him a fistful of pennies that had been collected by the families.

Jarek didn't hesitate, not for a moment. He started to climb over the fence.

"Jerry, where are you going?" Damian called.

"I have to go," Jarek called back as he straddled the top of the fence.

"You have your assembly in a few hours."

"I'll be back. Tell everyone I'll be back. I have to go."

Jarek exited the cab while Anka was still counting pennies to pay the hack. A crowd of women and children were gathered outside the tunnel's entrance. A group of sandhogs made a line separating the families from the angry police in their wool coats and riot helmets.

"There's the boy," one of the men called out.

Some of Jarek's father's drinking buddies grabbed his hands and dragged him through the crowd to "Old Man" Radaj. Radaj had been a sandhog for three decades and was now a shift manager. He knelt down next to Jarek.

"You can read, boy?"

"Where is my papa?"

"Can you read, boy?" the old man said shaking Jarek.

"Yes. Yes I can."

"Good. You should be able to squeeze through. There is a new machine in there gone berserk. Out of control. It caused a collapse, killed Jerzy who was driving the mad thing. Crushed the life out of him. The men made a hole.

You could squeeze through," the old man said. He turned his attention to a group of teenagers moving towards the entrance with a wooden crate.

"Wait here," he said, before pushing his way through to the boys.

"We're going to blow that thing to hell and get my brother and the others out of there," one of the lanky boys squealed when grabbed by Radaj.

Jarek heard some murmuring and then a great collective shout. When the shout ended there were moans and sounds of wood against bone. He turned and saw the sandhogs were in a desperate melee with the police. Both sides were swinging, clubs and pick handles raining down in a storm of pain and blood.

Jarek then noticed a group of worried-looking men in black woolen suits and top hats across the street. They were talking to the captain of the fighting police. Radaj pulled the crate of dynamite from the teen's hand. When it crashed to the floor, everyone, including the fighting men, stopped and held their breath. If it had exploded, only the men in the top hats would have been spared the carnage.

"The boss isn't goin' to let you destroy the badger, you idiots. It's worth more than all of you lot. And that's assuming you don't cave in the whole bleedin' tunnel and kill every soul down there. No! No, damn it! We're sending in the boy. Where is that damn book? Everyone just wait here for a god-damned second."

"Come here," Radaj said, pulling Jarek's arm roughly and dragging him past the still-fighting men to the gathering of top hats.

"Call off your bulls or you're going to get a proper riot," Radaj said.

A group of gang toughs were coming down 8th Street with clubs.

Jarek thought they might be the Plug Uglies but didn't know for sure. They wore derbies that looked like they were filled with rags for padding—that was the Plug Uglies' trademark fighting uniform. A man in a top hat and brushed out sideburns whispered something to the police captain, who blew his brass whistle.

It took more than a whistle to stop the brawl. The captain and a lieutenant had to enter the fray, pulling cops off the sandhogs. After a few tense moments the fight was over and both sides lined up again, staring, blood-covered and hateful.

"I will not have my machine damaged," the man with sideburns said to Radaj.

"Damn your machine, there are men, good men, stuck in there."

"They could be dead already," another top hat chimed in.

Jarek thought Radaj was about to hit the man but instead he turned back to the bewhiskered man in the top hat.

"This boy can read and understand things. He can also squeeze inside—his dad is in there. He can slide in and turn off that damned machine and the we'll dig the men out. All of them. It's the only way. We can't just wait for the machine to run out of coal. That could be an hour or more."

"Probably it will run out in a few minutes..." the associate of the top hat said.

"Shut your face. Just shut it. Have you ever been in there? What do you know about the tunnels?"

"I know the badger. I'm the engineer and I think I know more about it than a stupid–"

"Enough," the man in the side chops said with a wave of his silver-tipped walking stick. "Son, do you want to give it a try?"

"Yes," Jarek responded.

"Give him the book," Radaj said.

The engineer looked at his boss. The man in the top hat nodded and turned away to walk to his waiting carriage.

"Bones! Take the boy into the hole," Radaj yelled.

Bones was a tough Pole who had a number of torn shirt patches with copper police badges hanging from his belt. His front tooth hung like shutter missing a hinge. The thing that struck Jarek was how completely covered in dirt the man was—only his flayed knuckles revealed any color.

"You ever been in the guts before?" Bones asked prying open the gate to the IRT tunnel-site.

"No," Jarek said. Dusty darkness lay before him.

"Nothing to it. It's going to get hot and you can lose your sense. Just keep moving, always moving. When you get to the machine there will be air, humid as an armpit, but air. Breath with your nose, that keeps your lungs from cruddin' up as fast. What is that?" Bones asked, pinching the corduroy between his filthy and bloody fingers.

"Cour–"

Jarek's word were lost in shock as a cold bucket of water was poured on him. Bones held him tight as he tried to squirm away.

"Stop fidgetin'," the sandhog said, pouring a second bucket of water on the boy. "The guts get hot like a furnace. These soaked clothes are the only thing keeping from drying up like a dog shit in August. The hogs only work an hour at a time, and they're used to it but you... Well, just keep those clothes on. And put this on."

Bones pushed a metal flat-iron cap onto Jarek's head. It spun like a vaudevillian plate on a reed cane. The sandhog strapped the tin hat hard underneath Jarek's chin, the leather cutting into the boy's skin.

"When I set you down in the dropsey, you're going to see a pipe running along the floor to the left. That's a baby line. There are little teats on it every ten or so paces. You can suck on it to get some air from the upsides. Don't linger too much because the heat will get you. Take only when you get fuzzy in the eyes. Things start closing up on you, that's your brain dying. Got it? Keep moving and only suck when your sight closes up on you," Bones instructed, illustrating his point by looking through a closing fist. "Your pa is in there, eh? Well, so is my son, oldest. I know you want him back but if you can't do it just get back to the dropsey as fast as you can and Bones will yank you upsides. There are no heros in the guts. Only the living and the dead… don't get dead."

Bones gently guided Jarek onto a wobbly steel platform. There were worn leather straps hanging from the top of the cage. Bones pointed to the straps and Jarek stretched to reach one. The dropsey slipped and started to descend in fits and starts. Coal smoke rose up from the unseen boiler of the lift, and with each moment he felt the temperature rise. He wondered what folly it was that made people think that regular folks would descend so far under the city to ride a train a few blocks when they could walk in light and air.

"Follow the tracks!"

Jarek could hear Bones calling down to him, yelling over the clanking metal and chains of the dropsey.

Jarek could feel the badger before he could see or hear it. The floor of the dusty tunnel shook and loose rocks slipped from the walls and ceiling. The kerosene lanterns swung by their rusted hooks, casting nauseating shadows in the tunnel. Jarek kept his eyes on the tracks, periodically looking to his left to make sure the baby line was still there.

"Over here, come quick!" someone yelled from up ahead.

Jarek saw nothing but a cloud of dust that strangled his lungs. He breathed through his nose but soon is nostrils filled with dirt and was forced to breathe through his teeth.

"I'm Sven," said a boy not much older than Jarek but nearly a foot taller and twice his weight. They call me the Dane."

The floor was obstacle course of drills, picks, shovels and other tools he could not easily identify in the dust cloud.

"Can't dig much further or the whole thing may collapse. I'll give you a push, see if you can make it through," the Dane said, pushing Jarek into a rough-hewn crawlspace that the ten had carved out of the cave-in. "You've got to hurry."

Jarek wormed his way around obstacles of unseen rocks and debris. The suffocatingly close walls were slimy with humidity, turning the dirt into a slippery ooze. There were times he was crawling nearly vertically up and other times he had to brace himself with his palms to keep from slipping headfirst down into the dark. He wished for the baby line but there were no such comforts in this impromptu tunnel. He couldn't tell if he was suffocating or not, if his vision was narrowing or not, because there was nothing to see in the crawlspace.

Then he heard it. It whined and snarled in the darkness, crashing like some blind prehistoric beast in its death throes. Jarek was paralyzed with fear as he felt a warm trickle of urine soak his pants. Only the need for air forced him forward, overriding his body's terror. He dropped down into a larger tunnel, so hot that tears formed in his eyes. He scratched at the chin strap to remove the helmet so he could open his mouth as wide as possible to get some air. After an interminable coughing fit, Jarek saw the rear of it by one of the few unbroken lanterns—the badger. It blocked the entire tunnel, crawling slowly forward.

Even Jarek, who knew nothing of engineering, could understand the magnificence of the machine. It was the size of a carriage and constructed of plated steel. The rivet heads crisscrossing its body were each the size of his fist. Countless wheels struggled to keep it tethered to the track. Two large boulders had fallen from the ceiling and crushed the steel mesh of the operator's cage inwards, folding it like a discarded paper cup. Jarek saw the driver crushed under what was supposed to be a protective shield, his body wet and lifeless on the back platform.

Jarek wriggled his way onto the platform through a split in the seam of the cage. The operator's panel was covered with glass gauges, looking like a fly's multi-lensed eye. He had to intertwine his fingers in the broken cage to keep from being thrown off by the bucking engine.

Inside the pilot's cage he saw a periscope, like the kind at Coney Island, he thought, where for a nickel you could look inside and see the Statue of Liberty. Through the lens he saw the front of the mechanical beast, where dozens of sandhogs were backed up against a stone wall. They were naked, wielding shovels and picks. Some were digging furiously at the opposite wall while others, including his Jarek's father, were attacking the machine with pry bars. The dirty naked men reminded Jarek of cave men fighting a giant mammoth. He could see the men's mouths move, but couldn't hear them over the thunderous roar of the machine.

He pulled open the engineer's book. It was full of diagrams and numbers. Jarek had to hold the book close to his eyes because his field of vision had narrowed. There was not much time for him, and even less for the men who tried in vain to hold the machine at bay. He wanted to look back through the scope, but kept his eyes on the pages, searching for a way to kill this steam-belching beast.

He didn't understand everything, or even most of what lay before him on the page. It was complicated, and his overheated and oxygen-deprived mind was failing to make connections, draw conclusions. He gleaned what he could and tossed the book aside. He turned two gauges all the way to the left and then unlocked one of the battery of polished ebony levers. He took a breath and pulled the lever down, leveraging all of his weight to make it descend. The badger rocked back and forth and then there was a terrible screech like a storybook witch as the steam exhausted. When it finally died out, the machine was still. Jarek laid his head on the body of the dead driver. His eyes closed as he heard the shouts of joy from the sandhogs on the other side.

It took the men twenty minutes to break enough of the machine with their tools for them to squeeze through. Jarek's father took the unconscious boy and forced his mouth on the baby line while the other men started to frantically dig themselves out.

The police and managers had left by the time Jarek emerged with the surviving sandhogs. A couple dozen were alive and about that many had been killed when the badger had begun its rampage. No one ever determined if it was a cave-in that caused the machine to go wild without the driver or if the machine had gone astray and caused the collapse. Radaj collected money for a cab and someone donated a pair of clean trousers for Jarek. His clothes were long dried by the heat but heavily stained from sweat and grime. His white jacket was covered in dirt and the driver's blood.

The cab let Jarek out at Bentham Academy and he climbed the stairs and started down the hall. He never got to the assembly because Mr. Emmett-Hodges saw him first. He was holding his pocket watch open when he ran to the boy.

"You are almost late. It is five to–" the Settlement director's words stopped as he looked at Jarek.

Jarek tried to explain.

"You've thrown it all away. The scholarship, the money to the school. The project! Look at you. You look like a savage and smell like a sewer. If you go in there, you'll just reinforce what they believe about your neighborhood, that it's a savage place without a shred of the nobler stuff. Mr. Frick is up there now… do you hear him boy?"

The enraged director twisted Jarek's arm. Jarek could hear a man droning on to the invisible assembly. The school principal came out of the assembly.

"Go home. Go back to your damn neighborhood," the director said pulling off the lapel pin.

Jarek watched for a moment as the director and the principal talked nervously in the hall. He took off his jacket and left it on the steps. At least he still had his books, he decided as he walked home.

"To look at a machine?" Mathilda whispered to Pip who sat shame-faced on the bench in the bonds room.

Calamity, now properly medicated, was signing forms that secured Pip's release and formalized assurances that he would return to his court date to answer to the charge of criminal trespass. Everyone, including the rat-faced clerk, knew Pip would never return voluntarily.

"It's more than a machine. It's the future, Mathilda," Pip said, before Neal cuffed him upside the head.

"The future is people, not machines," Mathilda said.

"Come my love," Calamity called, "our evening's work is complete. The young Pip is as free as the wind."

The three started down the stairs of the courthouse when a bloodied group of riot police started up the stairs.

"Piss-it-all, see what you made me miss, for your damn machines?" Neal asked, looking longingly at the beaten cops.

"What the hell you lot lookin' at?" a police man with his badge torn off from his shirt said, stepping in front of the grinning Neal.

"I know you," the police officer said, staring hard at the group, "All of you holed up on Delancy Street, yeah I know your lot. You're some kind of nutters. Terrorists I heard."

Mathilda went up to Neal and steered him reluctantly away from the confrontation. Calamity waited until the others were down the stairs and hailing a two-cent hack before turning to the beaten policeman.

"Terrorists? No, we are something far more dangerous. We are musicians and we are the future."

This story, along with another four Seasonals and an introduction written by the Catastrophone Orchestra, was collected into a book called Catastrophone Orchestra: A Collection Of Works *published in 2011 by this magazine's publisher, Combution Books. Three of those seasonals have appeared in the pages of* SteamPunk Magazine *in the past.*

GEAROTICA

AN INTERVIEW WITH SHANNA GERMAIN

Illustration by Manny Aguilera

***STEAMPUNK MAGAZINE:** First of all, for those who aren't familiar with your work, can you introduce yourself?*

SHANNA GERMAIN: Certainly! I would try to curtsey or bow, but my balance isn't that hot and I've discovered a good introduction rarely starts with me accidentally showing my Lord of the Rings ("speak friend and enter") undies to the crowd. So I'll just say: I'm Shanna Germain. Writer, editor, leximaven, wanderluster, and geek. If it has something to do with words, I've probably written it. If it has to do with sexy words, I've definitely written it.

I like writing about things that "go bump in the night." Thus, my two favorite genres are horror and erotica. You can see my work in places like *Absinthe Literary Review*, *Best American Erotica*, *Best Bondage Erotica*, *Best Gay Romance*, *Best Lesbian Erotica*, *Blood Fruit*, *Queered Gothic*... I'm kind of all over.

***SPM:** What about steampunk erotica? What are the tropes that are developing? What can be hot about it?*

SHANNA: Steampunk erotica is really an interesting sub- or cross-genre. In a lot of ways, it's bringing new writers into the erotica fold, because we're seeing steampunk-specific writers say, "Oh, steampunk with sex! I could write about that." It's as though they've been given permission to enter a genre that they might have seen as taboo.

Steampunk erotica is also really fun and unique because you can use toys and spaces that haven't been invented yet. Vibrating steam machines. Hot robots with key-crank hardons. Airship sex. Leather-corset bondage outfits.

Of course, there has to be that other intangible element beyond the aesthetics in order to make it truly steampunk, right? And I know there are a lot of conversations about "what is steampunk?" and "what must fiction have in order to be classified as steampunk?" and those are very valid conversations. But erotica writers often seem to just "get it"—they're able to capture that slightly anarchist, slightly DIY, slightly against-the-grain, slightly dirty and gritty elements of steampunk in their stories, and I think that might be because they're used to writing about sex, which has a lot of those same sensibilities.

***SPM:** You've written a bit about how erotica writers are quite maligned. Yet you choose to write both erotic and non-erotic pieces under the same name, which is somewhat taboo?*

SHANNA: I think it's becoming less taboo than it used to be, and that actually makes me really happy. I love the work of early erotic writers like Sappho and Anonymous, but how much more could we have learned about their cultures and their experiences if we knew more about who they were?

I started under my own name out of a combination of pride and hard-headedness. The pride was: I'm putting all this work into writing this story, I'm claiming it, damn it! The hard-headedness was: No one can tell me that I'm not allowed to put my name on this, damn it!

As it evolved, my reasons for writing under my name became more complex. I felt that if readers had the balls to go to a bookstore and buy my books, I owed it to them to claim the work as my own and to stand behind the power of erotica.

I also quickly became tired of the shame surrounding sex—it was so prevelant, so strong and so full of vitriol. I realized that I could cower in the corner, I could rail against it, *or* I could present sex and sex-writing as normal, positive, pleasurable parts of people's lives (given the parameters of consent, safer sex, and all of those important elements, of course). The last one felt right for me—and it's amazing how many people have said, "Wow, I wish I could do that" or have asked me questions about sex because they felt like I was one of the few people they could talk to about it in a way that wasn't judgmental or shameful.

On the other hand, I totally understand that people have very valid reasons for writing under a pen name, and I respect that completely. It's definitely not for everyone. I feel lucky that I can openly be who I am, even if it occasionally means shouldering some of the negatives that come with that—some of the worst experiences I've had have been from women who stalk my blog, trying to shame me and berate me for writing openly about sex. Which actually makes me really sad for them, because you just know that something's not right in their lives. I want to send them a sweet "it will get better" note—along with a really fantastic vibrator!

***SPM:** What are some assumptions people seem to make about you as a person based on your work?*

SHANNA: People tend to assume that I've done everything I've written about. Which is odd, because I doubt anyone assumes Jeff Linsday (who wrote Dexter) is a serial killer or that Stephen King actually staked a bunch of vampires. Most erotica writers I know live very quiet lives on the outside. I always say I'm only a bad girl on paper (I don't know if that's entirely true, but it makes for a better quote than the truth, which is that I'm *mostly* only a bad girl on paper). Of course, I would love to have a life where I research sex experiences all day long, but then I'd never get anything written!

Sometimes people assume I'm going to have sex with them because of what I write, or they think they know my sexual inclinations because they read a story of mine once. In truth, all of my stories carry something of myself, but we're all so multifaceted that a small piece of me is just a tiny fragment of who I am in whole.

The other side of that is that in person, I look totally sweet and innocent. I'm the consummate blonde-haired, blue-eyed girl with the bounce and the smile. So sometimes when people meet me, they assume that I am only that thing, and then they find out what I write, and you can almost see them readjusting their first impression. Which is kind of fun, and also makes for some great conversations!

Shanna can be found online at SHANNAGERMAIN.COM

Manny Aguilera
MannyCartoon.com

A Note on the Legality of Poppies and Their Extracts:
The growth and use of opium poppies technically violates US drug laws, despite the potential for medical use. While growing poppies for ornamental use is often overlooked by law enforcement, any sign of intent to harvest is treated as a felony. For this reason any fictional character making laudanum would want to keep a low profile while growing, harvesting, or using opium poppies.

HOW TO (HAVE YOUR FICTIONAL CHARACTERS) MAKE LAUDANUM

by Canis Latrans
Illustration by Allison Healy

Laudanum has historically been used as an analgesic (pain-killer), soporific (sleep aid), and antitussive (cough suppressant). The ability of the average person to treat illnesses at home continues to decrease, while the cost of professional medical care rises beyond our reach. It seems likely that characters in our steampunk fiction might be interested in growing their own poppies and making their own laudanum.

Growing the Poppies

Many species of poppies exist, but *Papaver somniferum* contains the highest amount of active alkaloids. Try seed catalogs (often listed as *Papaver spp.* for legal ass-coverage) or the internet. Some say that the bulk poppy-seed available in health food stores will work, if one gets desperate.

A cold snap increases Papaver spp. germination rates, so once you've obtained seed find a way to get them cold. Some sow the seeds on top of the last winter snow, but one can also store seeds in the fridge for a few days to give them the necessary cold cycle. Once chilled, direct sow by casting seed thinly across a lightly raked garden bed. Mixing seed with sand can help get even coverage. Water gently, making sure the seed doesn't wash off the bed. Germination takes up to 14 days, so continue to water regularly. Ideally the bed will stay damp until the seedlings begin to establish roots. Once your seedlings emerge, water deeply but less often to encourage healthy roots.

Papaver seedlings have lance-shaped serrated leaves and grow quickly in good conditions. If aphids appear, spray the leaves with soapy water—strung-out aphids have been known to steal bicycles. After a few months, the plants develop flower heads, which will open sequentially into showy, easily-recognizable blossoms. The petals will drop within days, leaving seed pods which continue to grow for weeks. Let the first pod ripen fully, noting the size it achieves before it begins to turn yellow. This pod will provide seed for next year's crop and can be harvested once the pod rattles when tapped. Harvest the other pods when they reach maximum size but before they begin to yellow.

Ingredients:

- Dark glass container with airtight seal
- Alcohol, at least 80-proof
- 4 parts chopped poppy pods
- *1 part whole cloves*
- *1 part cinnamon*
- *1 part saffron*

(These last three were often used in traditional recipes, along with other ingredients from "nutmeg" to "ground unicorn horn." One can make a great tincture with only poppies and alcohol, but using these adds romanticism and flavor.)

Preparing One's Hypothetical Harvest

If using freshly harvested pods, carefully slice into the pod and remove the seeds, retaining as much of the latex, or sap, as possible. Chop the pods into small pieces. If using dried pods, grind in a clean coffee grinder (don't forget to clean it afterward). Combine poppies with any other plant ingredients in the glass container. Fill with alcohol and seal tightly. Leave the jar in a dark, cool place for 2–4 weeks, shaking it every few days. After it's soaked, strain the resulting tincture through a clean cloth into the clearly labeled tincture bottle.

The Dangers

- While it's homemade from plants, laudanum contains many powerful alkaloids, including morphine and codeine. This highly addictive tincture is also dangerous! Carefully label the containers with "POISON" and keep out of the reach in the uninformed, pets, and children.
- Test the strength by taking a gradually increased dose and noting its effects. Start off with a few drops, then wait 1 hour before taking a few more. While it may be a romantic way to go, resist the urge to take shots. It's entirely possible to cause permanent loss of brain cells or death from overdose.

Manny Aguilera
MannyCartoon.com

DAMN MY BLOOD

A CRITICAL LOOK AT SKY PIRATES

by Mikael Ivan Eriksson
Illustration by Manny Aguilera

Damn my blood...

You are a sneaking puppy, and so are all those who will submit to be governed by laws which rich men have made for their own security, for the cowardly whelps have not the courage otherwise to defend what they get by their knavery...

They vilify us, the scoundrels, when there is only this difference: they rob the poor under the cover of law, forsooth, and we plunder the rich under the protection of our own courage. Had you not better [be] one of us than sneak after the arses of these villains for employment? I am a free prince and I have as much authority to make war on the whole world as he who has a hundred sail at sea...

But there is no reasoning with such snivelling puppies, who allow superiors to kick them about deck at pleasure and pin their faith upon a pimp of a parson, a squab who neither practices nor believes what he puts upon the chuckle-headed fools he preaches to.

—Capt. Black Sam Bellamy

Airship pirates, we've all seen them. They are steampunk icons, but I have a problem with them. Well not really with them, but with the way many portray them, whether written, painted, built in 1:72 scale, or dressed up at a Con—their weapons, their equipment, and their means of transportation. All too often, something bothers me when I see them. And what is that, you ask? Well, to put it plainly: I don't believe them. Many say that the hardest thing is to make the fantastic seem probable; when fantasy writers describe orcs, dragons, and spells, they have to make them seem probable. The reader must be able to believe it could happen. The same thing should be true for steampunk—there has to be some kind of internally consistent logic to our creations unless we're making them for a Bugs Bunny cartoon. As a long time student of all things piratical, it annoys me so incredibly much that so few seem to have spent time researching what they are creating. I know that other steampunk icons suffer from the same abuse, but the pirate is the one closest to my heart, so it is his (or her) honour that I'll defend here.

The typical image of the pirate is an old one—the lone traveller who does not always see eye to eye with the law. We have seen him in many different forms: Han Solo, Mal Reynolds, Jake Cutter, and of course those from the "Golden Age of Piracy" like Jack Sparrow. They are more or less the same character placed in different settings. So when creating your own steampunk Airship Pirate, whether a character for a novel or a movie, a painting, or a costume, or making a prop, building a model pirate ship, or whatever you can come up with, please remember a few things. First of all, get to know the real thing. If you don't know squat about piracy you will probably not get it right. And I am not just talking about watching *Pirates of the Caribbean* and *Captain Blood* on DVD. I'm talking about research. Read some scholarly books. In the last few decades a number of really good historians have written a bunch of really good books on the subject. A short list follows this essay.

By now I can hear the first protests coming in: "It's just fantasy, how can you stand there and decide what is real in this fantasy world?" Well, the problem is that some things don't change. When it comes to pirates, some things have always been the same. It doesn't matter if you hook up with the Vitalienbrüdern in the 14th century, Black Sam Bellamy in the 18th century, or the Somali pirates today. They all use small craft, weaker than any military ship, with which they attack poorly armed ships, and they very seldom harm anyone they attack as long as they don't fight back. Things that are true for all other pirates would be just as true of airship pirates.

The Ship

The first thing that comes to mind that too many get all wrong is the means of transportation. Pirates have almost always had small, fast and highly manoeuvrable craft. The huge movie-ships with tons of cannons have very little to do with real life pirates. Bartholomew Roberts is about the only known pirate who used a big ship, if you don't count the modern pirates (and some fictional ones) that use a mothership as a base of operations. To find other examples you will have to investigate semi-legal privateers or military craft acting in a piratical way.

Another myth is that pirates used a lot of cannons. Well, yes and no. Cannons are a good thing to intimidate prizes with, but they were seldom actually fired. And if the were, they were only used to slow the prize down, or, in a worst case scenario, kill as many defenders as possible without damaging the cargo. On the high seas that meant pirates used grape- or chain-shot to get rid of the sails or sweep the deck. They never took a shot at the hull (as they do in almost every single pirate movie). When we talk cannons, there is also the question of defence. Sure, pirates could use cannons to defend themselves against attackers, but that was a last resort. Military ships have always been too tough for pirates, and pirates only fought them if there were no other way. They preferred to outrun their attackers in their light, shallow, and fast ships. One of the most popular ways Caribbean pirates escaped was to sail across shallow reefs in their shallow sloops, where the heavier deep-hulled military ships could not follow.

The Prey

Pirates used all their connections, knowledge, and cunning to find out where and when a defenceless merchant ship was going to transport its goods. Once they had that information they went on to get the prize. The usual way to do it was by sneaking up on them or by giving chase—only a good idea in areas poorly patrolled by military or other government agents. Once the prize was in sight, the pirate ship engaged it at the most difficult angle for the prize to defend; on the high seas that was from behind where few cannons were mounted. Once the pirates had gotten close enough, they had to make sure the prize didn't fight back. The best way to do that was to scare the living daylights out of them. This tactic worked best when the crew on the prize was aware that pirates almost never harmed crews that didn't fight back. This has been true throughout most of pirate history and still is true. Only when violence is used against pirates do they use violence themselves. It is an unnecessary risk and it makes it harder to get the next prize (if crews know that the best way to survive is to fight, they fight; if they know that the best way is not to fight, they will lay down their arms).

If the prize crew was a smart one, all the pirates had to do was get close enough to the prize to get over there and get the stuff. The usual way to do this was by pulling the craft close together with grappling hooks. When this was done, the looting began! But what if they put up a fight? Well the pirates fought back, and kicked the crap out of the prize crew until they gave up. Traditionally this meant sweeping the deck with grapeshot from the cannons and using snipers to get the opponents one by one. Merchant ships had small crews, so picking them off with rifles was actually a good idea. When resistance was weakened enough, the craft were pulled together with grappling hooks and both pirates and defenders prepared to fight in hand-to-hand combat. The preparation was usually to throw a grenade or two on the opposing deck to kill and maim a few more opponents before risking any of your own necks. Then they went in for the kill! The traditional weapons of choice were axes, pistols, and cutlasses, not sabres or rapiers. Why? Because a cutlass is short and a ship's deck is small and crowded with a lot of stuff everywhere in which a long blade could get stuck. A cutlass is also an easy weapon to use. Using a fancy rapier needs training, but the crude cutlass is taught in no time at all; you just have to chop your way through your opponents. This is true even today, though today the cutlass has been replaced by the machete which in many ways looks and works the same way as the cutlass. Since pirate crews have almost always been much bigger and better motivated than crews of merchant ships, the outcome of hand-to-hand combat is almost always obvious from the beginning; this also means that time spent fighting was pretty short once the boarding started.

The Loot

Pirates were usually pretty poor; they didn't swim around in gold and diamonds. They needed a lot of stuff to keep their vessels afloat and supplied, so when looting a prize they took whatever they needed—tools, food, sails, whatever was not nailed down and some things that were. Then they took the cargo, not because they needed it but because they needed the stuff they could buy from the money they got when selling it. Then there is the crew. First of all, don't kill them! Remember what we talked about earlier. Pirates wanted them to spread the word about pirates letting people survive if they don't fight. In the Golden Age Of Piracy, though, there was one exception—the ship's officers. When dealing with the officers, the pirates first of all considered whether they had forced the crew to fight the pirates. If they had, tough luck for them. Punishment was to be expected, and they were killed. If they had not forced the crew to fight the pirates, the crew was asked if the officers were good to the sailors or if they were utter bastards.

If they were bastards, tough luck for them. No more officers on that ship. If it turns out they were a decent bunch they were treated like the rest of the crew. This meant first of all that they were offered a chance to tag along with the pirates. If they didn't, no hard feelings. But the question of what to do with them now had to be dealt with. The usual ways were to let them go on their own ship, keep them on the pirate ship until the pirates could get them safe ashore, or let them go in a long boat with enough water and food to get themselves back safely to shore.

Once the pirates had the goods, they needed to sell them. There have always been two major ways of selling the loot. One is to sneak into a town with crooked merchants and other officials where you can buy what you need once you've sold the goods to the land-bound crooks. The other way is to go to a pirate haven and sell the stuff there. These places have not been so common throughout history, but they have existed. New Providence on Nassau is probably the most famous example. Once the nicked stuff was sold and what needed replacing on the ship was replaced, it was time to divide the plunder. A pirate did not get paid a set salary; as early as the Vitalienbrüdern, pirates used the system that each and every one got a share of the loot. By the 18th century this system was rather advanced and was one pirate tradition that actually can be interpreted as some kind of proto-socialism. Once the loot was divided and the pirates were safely on shore, the party started. The party continued until there was no more loot to spend, then it was back to sea again, and it all started all over with hunting for new prey.

The Pirate

Now that we know how the act of piracy was conducted, it is time to look at the pirate him/herself and how pirates ran their ships. First of all, most pirates were from humble backgrounds. Most were born into the lower class and had little or no reason to believe that their situation would change without piracy. Few fit the movie profile of the "gentleman pirate." The only known example from the Golden Age is Stede Bonnet, who, quite honestly, was a rather lousy pirate. Most pirates turned to piracy because they had few other choices. Inhabitants of Hispaniola became pirates when the Spanish killed all the game animals that the buccaneers had hunted, much as the Somali piracy started when fishermen were forced to defend themselves against the illegal fishing and waste dumping that had totally destroyed their livelihood. During the Golden Age piracy was a way to escape the harsh conditions crews endured on merchant ships, naval vessels, or, worst of all, slave ships. Very few started a career of piracy in search of adventure or for the fun of it, and the reason for that is simple—as Bartholomew Roberts put it, "it is a short life, but a merry one." A pirate's career was usually a short one with a violent and bloody end. Very few pirates lived to see retirement. Pirates like Israel Hands, who got shot in the leg and therefore retired from the Blackbeard's crew, were the lucky ones. If you look at the Somali pirates today, it's the same thing. What else does life in Somalia have to offer? Well, nothing, so you might as well risk getting killed by the US Navy to enjoy a short life with lots of khat, cars, and women instead of a short life with starvation, abuse, and misery.

During the Golden Age many pirates were former slaves or subject to indentured servitude, a kind of slavery that at least officially had a time limit, and the conditions for a labouring sailor were not much better. So most pirates were used to being ruled by absolute authority. This is probably why most pirate crews during the Golden Age signed what they called articles, an agreement on how to run their ship in such a way that no one would get abused and no one could take unfair advantage of another. The Golden Age pirates were, for their time, extremely democratic. They voted on almost everything and in no way was the captain in charge. Even the most feared and legendary captains had to follow the decisions of the crew. When Blackbeard and his crew seized Charleston he told the hostages they had captured that he would like to set them free, but the crew disagreed so they had to stay a wee bit longer. And this is probably what you get when you put a bunch of people who all hate authority in a constrained environment like a ship, and they have to make things work. They create democracy in the most anarchistic sense of the word!

Pirates around the world have always dressed practically when out on the sea. That means they looked more or less like any other sailor. This also went for women pirates. When Anne Bonney and Mary Read walked around on the deck of Rackham's ship they were dressed like the rest of the crew in trousers and shirts, according to witnesses. The were not disguised. According to witnesses, it was quite obvious they were women, "due to the largeness of the breasts," but were dressed like men because it was practical. But shore leave was quite another story. According to archaeological discoveries from Black Sam Bellamy's ship The Whydah and numerous contemporary witnesses, pirates really dressed up when they partied on land. In a society where dress was closely connected to class, they wore clothes from the upper classes and made them their own. They showed off big time, partly to spite those with power, but just as much because they simply wanted to look good.

The Steampunk Pirate

So where does this all leave us as we create our pirates? Well, if we follow the descriptions above, we start with a light and fast ship, perhaps a small submarine, or a dirigible where the gondola is based on the sloop of days gone by. But keep it fast and light. You might even want to try using small WWI-style biplanes to force down prizes and plunder them. But then you run into the problem of getting away with the goods, since biplanes hardly can carry any cargo at all. On the cannons I vote no. Go for rifles, or maybe harpoons to trap the prize with, but nothing that risks sinking the prize by damaging the hull (if sailing) or the envelope (if flying). Instead, target the propellers or whatever pushes the ship forward. The pirate ship would surface next to the prize, if using a submarine, or, if it all takes place mid-air, sneaking up from underneath, or perhaps, like the dog fighters of WWI, attacking with the sun at your back. One problem for an air pirate is how to let the crew go once you have seized your prize. How do you do that? Parachutes perhaps, but it might take a lot of parachutes to release an entire crew.

When creating your pirate, think of this—they are probably not looking back at a happy childhood on the family farm. Instead, they probably hate their crappy backgrounds and never want to go back. By turning pirate, they have nothing to lose and all to gain. They probably hate the rich and upper classes, who probably abused them when they were young. Pirates are likely to be far more democratic than their culture of origin, to be pissed off at society, to like to party, and to dislike violence.

Dress your pirates in a way that fits the vessel they is using. If using a submarine, base the look on 19th-century sailors with striped shirts and white trousers. If using a dirigible with a gondola that looks like a sailing ship, mix it up between the classical Robert Newton-styled 18th century pirate and the 19th-century sailor. If your vessel is really fast, like a biplane, and open, a leather aviator cap, goggles, and a warm jacket will protect your pirate from the cold winds (think Manfred von Richthofen); if the plane is enclosed you can use the more casual "captain's hat" in the way Jack Cutter and Johnny Hazard do, but remember the top should be soft, otherwise you won't be able to get the headgear over it.

If you follow these pointers, and most of all the one about studying the subject, I think paintings, stories, portrayals, and miniatures in the steampunk-pirate genre will be so much better. The pirates will come to life in a way we have never seen before. And in the end, we will all be much more satisfied with our creations.

Books to Get You Started:

Scholarly books:

David Cordingly: *Under the Black Flag*

Angus Konstam: *The History of Pirates*

Gabriel Kuhn: *Life Under the Jolly Roger*

Marcus Rediker: *Villains Of All Nations: Atlantic Pirates in the Golden Age*

The original sources:

Alexander O. Exquemelin (in some editions John Exquemelin): *De Americaensche Zee-Roovers* [*The Buccaneers of America*]

Capt. Charles Johnson (in some editions Daniel Defoe): *A General History of the Robberies and Murders of the Most Notorious Pirates*

For the legend:

Howard Pyle: *Howard Pyle's Book of Pirates: Fiction, Fact and Fancy*

Robert Louis Stevensson: *Treasure Island*

TAKE TO THE WIND!
AN INTERVIEW WITH GREG RUCKA

I was lucky enough to sit down with Greg Rucka, writer and creator of the new steampunk Webcomic, Lady Sabre and the Pirates of of the Ineffable Aether *which is currently free online and updates every Monday and Thursday on* WWW.INEFFABLEAETHER.COM.

by Juan Navarro

JUAN NAVARRO: *What is* Lady Sabre and the Pirates of of the Ineffable Aether*?*

GREG RUCKA: It's a steampunk action/adventure serial that Rick Burchett and I are doing as a webcomic. Or, another way to put it, it's our opportunity to indulge our love of several different genres in one spectacular, limitless setting.

JUAN: *With all the different projects, how did this web comic come to fruition?*

GREG: Rick and I have been trying to work together for literally years, now, only to see project after project shot down in one form or another. We'd been working on developing another piece altogether, when suddenly we both seemed to have the same realization—that if we *really* wanted to be working together, we should just do it ourselves, put it out on the web. As soon as we thought that, it was literally a very short hop before we had the concept for Lady Sabre.

JUAN: *So this was a bit of a DIY exercise then for your guys too? What do you think about the webcomics market and the phenomenon of just getting your book out there through the web?*

GREG: I love the opportunities that webcomics provide—the ability to self-publish our material is incredibly liberating, so there's that. As to the market, per se, I don't know nearly enough to speak with any kind of authority on the economics or it. The fact of the matter is that Lady Sabre is a labor of love right now, and my suspicion is that we've long passed the point of making serious bank on webcomics, at least for the time being. But that's okay—we're able to put our story out there for the larger world to see, and hopefully, as more and more people come to the site, and as more and more people seek to support what we're doing, that'll turn around. It's a wonderfully level playing field, and I feel very strongly that our success or failure is, in large measure, in our own hands.

JUAN: *Why steampunk? Or better, what is steampunk to you or within the world of Lady Sabre?*

GREG: Rick and I have talked about this a lot, actually, because it wasn't as overt as saying, hey, steampunk! But so many of the pulp and genre elements we wanted to play with are effortlessly at home in steampunk, it really became a no-brainer; the idea of taking an era and style that appealed to both of us, and then running full on Jules Verne with it.

There's more to it, of course. The design opportunities are limitless, and that's certainly a huge factor for Rick, but for myself, I love the feel of the era that never existed, and the freedom to be found within it. Our world for these stories isn't supposed to be Earth, but rather an Earth-analogue. Telling a story about the rise of technology, as it collides with a decline of magic, in an age of exploration… these are all steampunk hallmarks to me.

JUAN: *What can people hope to see in the future for* Lady Sabre and the Pirates of of the Ineffable Aether*?*

GREG: We've just finished with Chapter One, and are now into Chapter Two. So one of the first things people can count on is better storytelling! Seriously, my pacing for the first Chapter stank; it reads fine all together, but as a bi-weekly post, it was sorely lacking. This is still very new to me, but I'm optimistic that we've worked out some of the most glaring problems. So there's that.

Chapter Two introduces a new Land—Tanitin—and two new characters who'll factor crucially into this first story. Chapter Three brings us back to Lady S, aboard the good ship Pegasus, as she struggles with a new problem, and then… we get to meet our Bad Guys.

So, yes, there's plenty to come!

JUAN: *You've mentioned RPGs and their influence and help in writing, so maybe we can see a Lady Sabre roleplaying game, hmmmmm?*

GREG: Ah, I'm not a master of game mechanics, sadly! That said, someone wants to talk home-brew, I'm all for it! Part of what I'd love for the comic is to see it proliferate amongst the audience, you know? If we can create a world that is compelling enough that others wish to set stories within it, to explore it further, that's possibly the highest compliment Rick and I can earn!

JUAN: *Awesome. So one final question: if you could have any invention from* Lady Sabre and the Pirates of of the Ineffable Aether *or from steampunk in general, in hand, what would it be? Or what do you wish ran on steam?*

GREG: Oh, I've got to stick with Lady S for this answer; it's either her smallsword, or the Pegasus itself. Though parking the Pegasus might be difficult, and I'm not sure I could scrape together the requisite crew. But I love sword fights, and her blade is kinda awesome, in my opinion.

An essay by Katherine Casey
Illustration by Tina Black

FRANCES WILLARD

SUFFRAGIST, TEMPERANCE ADVOCATE, LESBIAN

Eventually, Frances Willard would be known as the president of the Women's Christian Temperance Union, an organization responsible for mobilizing thousands of women across the country to fight the evil of alcohol and lobby for Prohibition. The university I attend would name a dorm in her honor 50 years after this accomplishment, thus starting a long and noble tradition of naming buildings after uptight, moralizing women, and a copy of her diary would end up in our library. It sits between the books on treating alcoholism and on alcohol in American culture.

So I was surprised to read it and find that, at the time she was writing her diary, Frances Willard wasn't the anti-drinking political mastermind we'd learned about in class. She was 19 at its start, about to graduate with a difficult degree from North Western Female College (despite her mother's insistence that she only need learn the basics). As the diary unfolded she came up against the problem, time and again, that the people she fell in love with, the way she felt a woman ought to love her husband, were always female. In her diaries she depicted her movement from confusion, denial and self-criticism to accepting herself and planning a future for herself as a woman who loved women.

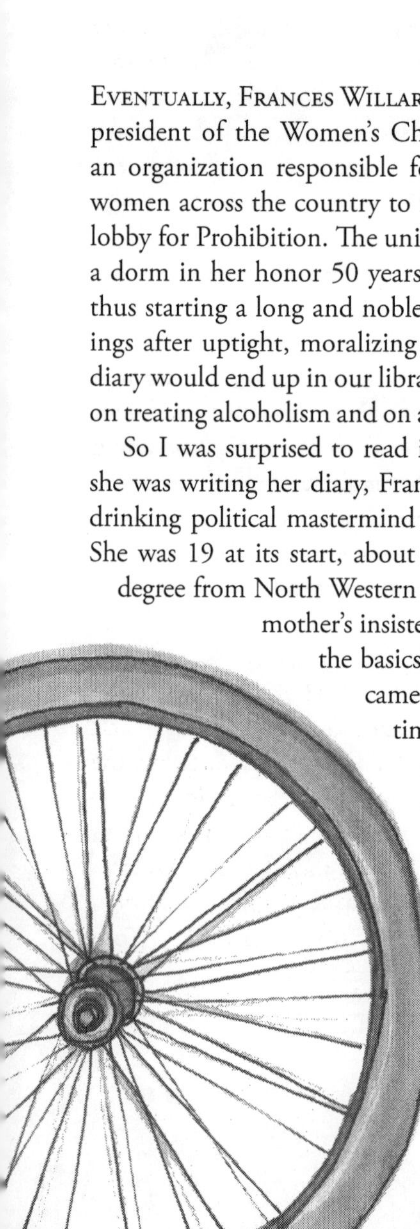

For most of the seven years or so that the diary spans, Frances was in love with her beautiful neighbor and long-time friend Mary Bannister. They had what has been called a "romantic friendship," a common thing in a time when so much of life happened only with others of one's own gender. They cuddled in bed, exchanged endearment-dripping letters, even shared their diaries with each other. Frances's was full of page after enraptured page about how wonderful Mary was, and Mary seemed to have returned her friend's passionate affection.

However, things changed quickly—Mary became engaged to Frances's brother, Oliver, and suddenly relations between Frances and Mary were expected to become cordial and sisterly overnight. Mary seemed to have had no problem making the change, but for Frances, it was agonizing. Mary's engagement marked the start of a horrible six months. Frances would be cold to Mary, write a passionate apology declaring her lasting love for her friend and her intention to behave properly towards Mary, then fall into despair at her "unnatural" affections and start avoiding her new sister-in-law all over again. Meanwhile, she became engaged herself—to Charles Fowler, a kind, intelligent seminary student who respected Frances for her intellect.

Reading her journal was like reading a soap opera, with a nice healthy dash of teen angst. Did she love Charles, or did she not? She said she did: "He loves me very much, I know, and my whole life shall prove to him what I say so quietly in words, that *I love him* as a woman can who has not *loved any one* before…" she swooned, but cringed at his kisses. Slowly, she realized she had loved someone before—*Mary*, whose name she underlined every time it appears in the journal. When she and her new sister-in-law had a falling out, it only strained her relationship with Charles further. When he visited Frances and asked her why she was so unhappy, she admitted that she loved Mary more than him. At first they agreed that it didn't matter, since she continued to say she loved him. However, a month later, as they walked together near Lake Michigan, she told him that she did not love him, but agreed to do as he wished regarding the marriage, and he decided to stay engaged to her.

The weeks that followed were composed of endless serious conversations: with her mother, who thought her crazy to even be thinking that she loved a woman more than her fiancé; with her father, who supported her, though she does not record his thoughts on her sexuality, if she told him at all; and of course, with Charles. They made all sorts of agreements—there needn't be physical demonstrations of affection in their marriage, no consummation until she was *sure* (her italics) one way or another about her feelings for him. But she still felt "tormented with the abnormal love & longing of a woman *for* a woman" by November 1864, though she wrote in December that she did indeed love Charles and still intended to marry him. By January she had changed her mind yet again (and Charles, poor man, had gotten tired of this nonsense and was starting to reconsider as well). In early February, they exchanged a series of affectionate and cordial letters and decided to pretend that whole engagement thing never happened. She'd always referred to him in her journals as "Charlie," but by the next entry he was "Mr. Fowler" and that was the end of that.

Eventually, after many long, angst-ridden letters and heartfelt conversations between Frances, Mary and Charles, Frances got over Mary. She almost at once had a new name to underline in her journal—*Ada*. By then she had started teaching at a women's college, and seems to have come to terms with her strange situation. "I like to think how sweet it is for women to whom God grants the sweet hallucination of '*la grande passion.*' I know how it might be—to feel one's self *clinging* instead of standing upright or having other's cling to one as my sweet Ada does to me." She described Ada as her wife, and worried that someday Ada would want to go on and marry a man as Mary had.

I suppose Frances would probably be offended to hear this, but her heartbreaking series of relationships as a young woman just older than myself reminded me of high school. Her striking love life only serves as background to what's even more surprising and beautiful in her journals: the way she slowly started describing her own sexuality and decided that it needn't prevent her from having the life she wants. She wrote in words that anyone coming out now would understand. "It is strange—incomprehensible save to myself & God who made me as he did," she wrote, around the beginning of her falling out with Mary. "My nature is what makes [my family] sad, and that I can not help." As someone thoroughly acquainted with having to explain my nature to God and my family despite not being able to help it, I read Frances's journal with a thrill of recognition.

About half a year later, with her engagement ended and her relationship with Mary thoroughly ruined, she wrote that while she thought she had "missed the greatest good of a women's life" (by which she means marriage), "there are a thousand pleasant, worthy things besides *that*.... There is good to be done—knowledge to be acquired—Friends to be made and kept..." She added that "Naturally I *love* women and ... can feel no earnest, vigorous love towards *their brethren!*" Her determination to make a good life for herself despite giving up on that most important part of a Victorian lady's life takes up about a page and a half of exclamation points and dramatic italics, and leaves the reader wanting to stand up and applaud at the end, or perhaps hug her for accepting herself a century before queer pride was ever an idea.

All of this was long before her involvement with politics—though the journal was shelved with the alcohol books, she never mentioned the evils of it that she'd later make her career enumerating. I find myself wondering if her conviction that women could have good lives outside of their traditional roles, brought about as she came to terms with her sexuality, influenced her later campaign strategy as she encouraged women to "do everything" and take active roles in politics. I still don't think the Willard who my school chose to name a dorm after and I would get along, but the young woman in her diary, falling in and out of love and dreaming up lofty goals despite it—she's someone I would want to ask out for coffee.

Lichens

by Pinche
Translated from Italian by reginazabo
Illustration by Kevin Petty

First Installment

While she walked the tip of her shoes confronted the debris of crushed bricks and dull metal objects.

The air was biting cold; she adjusted the collar of her coat to her neck.

Just ten minutes before a relentless rain had been dropping, but now the sky seemed at peace and the air cleaned up by the weeping-like outburst.

She sniffed the air—all around the ruins smelled like forests. The broken bricks, the splintered concrete blocks, those small asphalt islets that still emerged from beneath the grass.

She got to the Mouthed Gate. It bore that name because what remained of its two iron doors was a pair of large slivers at the top, forming the cheekbones of an open-mouthed face.

It was a marvellous gate, one of her favorite, one of the many that still guarded imaginary palaces and invisible factories. Like the rest of them, the Mouthed Gate was totally useless. It was a mere memory of what it had once protected. And like the rest of them, you would have never dreamt of not using it, of mocking it by going around it.

Sitting on the banister of an orphan window there was Typtri.

"Hallo, Zam."

"Hi... today there was this slow rain falling instead of the usual *neige*... have you heard?"

"Mm. Nice. I like rain. Afterwards smells are sharper."

As Zam got closer, her gaze wandered behind a long line of ants that ran along the banister, each one with its tiny load of debris.

"This place will soon be eaten, just like everything else. I'll be sorry when the Mouthed Gate won't exist anymore."

"Zam, perhaps I have found a place where we could look for your fuse. Down at the bay I heard of an old factory in City22—it produced electronic and hydraulic components. They say that many walls are still standing—in some parts there's even a roof. I wouldn't wonder if we found something that is still usable there."

"It must have been looted..."

"Yeah, sure, but nobody would've wanted your fuse... it hasn't been used since the end of the 20th Century."

"Don't know, Typtri. City22 is far away. Sometimes I think that I should give up looking for that damned fuse. I'm just making you all waste a lot of time."

"Zam, do you seriously think I've got something better to do?" Typtri would have smiled, if it had been able to, and Zam appreciated this effort anyway.

"Besides, who knows? At the factory in City22 I might find some oil, or some gears..."

"Yeah, alright. But we must leave tonight—who knows when we'll have another chance of flying without the *neige* falling."

Typtri collapsed down from the banister—with those short legs, it shouldn't have acted so athletic, and with so much corroded iron around, Zam always feared it would crash into pieces at any moment. It was a funny device, Typtri was—its looks always contradicted its words.

"Let's meet at the bay in a couple of hours, then. I just need some time to get a couple of things. I will fetch some oil and water I found yesterday: it should be enough to get us to City22."

"Okay, Typtri. See you in two hours at the bay."

Two hours was a time span vaguely included between now and never. Measuring time had become too difficult for anybody to still believe it could be worthwhile. Keeping a clock going, even if it had been possible, would have meant using precious gears. Watches and alarm clocks had been cannibalized decades ago. Down at the bay they had a couple of sundials and they counted days, but it was more out of habit than for any particular reason. If you wanted to arrange a meeting with someone, you would approximate by remembering old conventions: a quarter of an hour is less than an hour, which is less than two days... Those who arrived first just waited: not that they had anything better to do anyway.

With years it was easier—seasons could still be distinguished pretty well, despite the *neige* that fell continuously to remind everyone of the days of the bombs.

Zam spent her imaginary couple of hours going home to fetch her backpack. She filled it with her pilot cap, the white umbrella, a high neck sweater, knickknacks, and various spare parts. She had not travelled much lately, and resolved to celebrate the event by wearing the green striped dress she had found so many years ago in that almost-intact theatre. It made her feel like a porcelain doll.

She tied her hair with a big red satin ribbon. With her blue hair and the green dress, she was already wearing more colors than she was used to. She settled the question with her black rubber boots, took her backpack, got down from the tree, and walked towards the bay.

The bay was in turmoil, as usual. Emu and Park were tinkering around some golden pistons mounted into a plastic tube that was as big as a fridge, the Runts were sitting on the ground watching, and every now and then they laughed and threw a small stone at Emu.

"Hi, Emu. What are you doing?"

"Hey, Zam. This is the new machine I was telling you about the other day. Well, this is a piece of its engine... I'm not sure, but I guess the distance between the pistons is correct... what do you think?"

"Well, yes. Have you tried it yet?"

"Oh, no: we've just finished fitting the pistons inside..."

"I can't give you a hand, Emu, I'm sorry... I'm meeting Typtri, we're going to City22."

"Because of the factory? It was me who told Typtri..."

Her big horns framed her face as a proud smile brightened up her face.

"Some days ago a guy passed by. He came from City22 and told us about this well preserved plant. He said that he found some perfectly-kept rubber reels. He uses them to insulate the implants he's got on his back..."

"Don't know, we'll see, Emu. I don't count on it very much anymore. It's a very old kind of fuse, when the first *neige* started falling it hadn't been used for decades..."

"We'll see. Ah, listen, Zam, while you're around there, see if you can find a few bolts, size 15. We don't have many of them left and we need them for the Runts' blowguns."

"Okay. I'll see if I can find some."

The bay embraced the lives of all the area's Remnants. When you got there, the whole view could make you dizzy. High above, dozens of trees hosted a network of homes, walkways, and slides. Down below, narrow paths of grass worn by steps meandered among rusty heaps of iron, contrivances, and parts of every shape. The bay rose over a huge mechanical laboratory.

Climbing the stairs that led to the Golden Bar, Zam met a Runt coming down. It was Hud, with his load of plastic bottles.

"Fifteen today. Zam, when you get down will you bring me the rest? I must have left four or five upstairs that didn't fit in my backpack..."

"How far have you gotten with the house? Have you finished the walls yet?"

"We're nearly finished... we need eighty bottles or so and then we can start with the roof. Berc's gone to tale a heavy load at the old slump. After that we should be done."

"It will be wonderful, Hud. I'm leaving, but I'll come see how things are going as soon as I'm back."

Take a plastic bottle, fill it with sand, and you've got a brick. Jam four bricks into each other, and you have started a wall.

The Golden Bar was almost empty. Typtri beeped to greet her while Zam got a glass of juice.

"How long will it take to City22, Typtri?"

"Not sure. If the flydrome catches speed and the *neige* doesn't fall, we should be there in a few hours."

Zam thought about her last journey, looking for that river in the West... It had taken her three days to get there, only to find that the river had dried up.

"Better than I thought."

At the garage they found Lip half inside one of the large flydrome tanks. When he emerged, his arms were covered with sludge to the elbow, his face was red, and his shirt was still too white.

His face brightened up.

"Zam! The flydrome?"

"Yes, Lip, we need your help to push it to the runway."

The flydrome Zam always used was made of brass. Emu had built it after ransacking a huge ocean liner that had got stranded on a beach up north. At the bay, the brass from the finishings of the ship had been enough last them years.

Zam liked it especially because the seats came from the same theatre where she had found her dress. The were lined in brown velvet, a bit worn out, but comfortable nonetheless.

They filled the water tank, greased the engine, and loaded the flydrome with their supplies.

When they were ready, while the engine started with a puff of steam, Zam turned to watch Lip, with his handsome black and white face, his grey apron, and his purple eyes.

What could her red satin ribbon possibly look like through his psychedelic vision?

Typtri climbed next to her, Zam put her cap and goggles on, and they took off.

"Zam, I think City22 is down there."

The flight had been refreshing. Zam had never gone so far north, nor had she ever flown without the *neige*. After some hours it had started again though, carrying along that creaking silence. Presently, as they got closer, a beautiful city made all its best to be noticed.

A river, almost drained, ran almost from one side to the other, cutting it in half. In the north, wide sweeps of shrivelled high-rise blocks looked like sleeping elephants. In the center, instead, grass had grown all over, and several trees were blossoming inside big, roofless gothic churches.

"Typtri, I've already seen this city… I mean it…"

"Well, of course you have, Zam. This is Paris."

They landed in a wide field, on the east side.

The cold stiffened her fingers, and Zam regretted leaving her gloves behind.

They made their way towards the area that had looked more intact from above.

Typtri had a funny way of proceeding, sniffing the air and stumbling in tires and rusted iron scraps.

He looked around. There were no more lichens. What remained was a strip of maple leaf he had nibbled the day before yesterday and left there to dry.

He made do with it, just to have something to munch on his way home.

In the last week the days had started to defrost. It was still cold, to be sure, but not as cold as before, not the freezing cold that bit into his bones.

While he went up the river, he stopped just a second to check that the sky wasn't threatening a storm. And while his gaze turned back to the ground, he noticed a glint of something hidden among the dry branches.

It was a bolt. He had seen more of them when he'd ventured to the big empty factory on the other side of the hill. As to how it had got here, it was hard to figure out. Perhaps a tired magpie had let it fall along its journey.

A nice bolt. A big, brass-colored one.

He moved it with his hoof so as to keep it hidden.

Zam watched as Typtri moved a concrete-framed glass block with its large tongs.

"If we had a larger aircraft, this large block would be very useful back at the bay."

"Typtri, a lot of things would be useful at the bay, but I'm getting a bit tired of moving things around the world."

She looked at her shoes. She was sleepy. That feeling had started to encrust on her skin since the AP 7080 fuse had broken.

They kept walking for a long time, kicking styrofoam boxes and making chocolate wrappers rustle on the ground.

A dump—she was deeply depressed of living in a huge meaningless dump. It was like living inside a smoked glass globe.

But something had started moving as it had never done before, not only because there were no alternatives. Zam had long acknowledged the irreducible hyperactivism of those who are building something, and that feeling was so beautiful it made her tremble.

But sleepiness caught her eyes at every moment.

That damned fuse. Imperfect heritage of an imperfect parent industry.

In front of his house door, he suddenly felt like sitting down. What was it like to sit down? He didn't remember anymore. It had been exciting and marvellous to be turned into a deer, but now and then a tear formed deep in the right corner of his left eye, a tear that contained lost memories. Trivial memories of trivial things. Imperfect heritage of an imperfect parent industry.

Zam stopped abruptly.

In front of them, at the end of the road, there was a wooden house, like the ones in fairy tales: a little house in the middle of a small forest. And in front of the house door there was a big deer squatting in a strange position, with one leg over the other. A sitting deer.

In his gaze there was the careful concentration that comes with an unsustainable effort, mixed with a distracted, bottomless sadness.

Second Installment

The deer stared at them with a perfectly bovine attitude—with watery eyes and a mouth curved in a predictable, peaceful smile.

When he got up, the effort warped the corners of his mouth and his long legs made many ungraceful movements as he attempted to find a normal position.

Zam kept gazing at that clumsy deer, recognizing something familiar, a well-known inadequacy.

Clangs of metal snapped their fingers inside Zam's mind: Typtri was already far away, lost again as it rummaged among the ruins that lined the forest. Zam reached it, trying to dodge the awkward feeling that the deer was following them. But in fact he was, slow and relentless as a cow chewing cloves.

"That must be the factory."

Typtri had stopped before a rust-eaten gate. Behind the gate there was one of those big crossing bars with red and white stripes. It was raised to let people through, and probably it had stood there for a long time. Ivy ran along the bar and the asphalt desperately looking for a lump of earth. It also ran along a hibiscus, a great purple hibiscus that kept a blooming guard over the gate.

Behind the bar and the gate, there was the factory. Three large, faded yellow sheds, their herringbone-shaped roofs turned westwards.

While they got into a hall, Zam tripped on a rusty beer can and the welcoming silence of the place caressed her cheeks.

For a long while she and Typtri picked up the bolts and nuts that were scattered all over like seeds of a strange metallic plantation. Then, as they walked through the narrow corridor linking the first shed to the second, they knew from noises and voices that the factory was not abandoned at all.

The second hall was much better lit than the first—the roof had partly collapsed and the *neige* gathered on the floor and the debris whitened the walls all around with its reverberation.

Underneath the part of the roof which had not crumbled, there were two women and a man, sitting around a large machine that rattled and creaked in an extraordinarily invasive way.

They were weaving long white cloths that were gradually piling up on the floor in loose rolls.

A fourth man went to meet them with a slight bow.

"Good morning, you are from the Ministry, I figure. Are you here for the inspection…?"

Zam waited for some vaguely logical words to come to her mind.

"No, look, actually we come from the Bay, north from City11, we were looking for spare parts. We've been told about this factory where we could find them… and anyway, what Ministry are you talking about?"

"Oh, no, I'm sorry, unfortunately we do not produce mechanical or electronic components, if that is what you are looking for… in fact we do not produce hydraulic components either…"

"Well, no, of course you don't produce them…"

Zam had started plunging into discomfort since that strange man, dressed up as a Far West post office clerk, had spoken his first words.

"Indeed, no. Our activity uniquely consists in flag manufacturing."

"Flag manufacturing?"

The conversation had definitely taken a surreal turn.

"Exactly. Come with me, I will show you the production chain. Unfortunately our firm is small, we had to do without most of our staff and productivity was considerably reduced…"

"Your firm?"

"Indeed. Oh, I must apologize… I have not introduced myself… I am Theodore Ri, project manager of Sic Corporate."

"What are the flags for?" Typtri said to stop Zam from continuously repeating the last few words the man spoke.

"Ah, well, right… we only operate under the supervision of our majority shareholder…"

"Yeah, but what are they for? They're huge… they must be ten meters long…"

"It is a standard size, if you have been sent by the Ministry, feel free to check personally that we abide to the rules of the circular letter number 748494… and to the regulation for workplace security as well…"

Zam gazed at her own feet. The man had suddenly turned red, his hands gesticulated anxiously, and she was afraid that by continuing to stare at him as though he was an alien she might end up making him burst into tears.

"No, look, we are not from the Ministry… and what Ministry anyway? There are no ministries anymore…

there is no state, no government… how can there be any ministry?"

"Well, yes, in our country there is a tense atmosphere, you are right. Truth is that the government has abandoned us… do you know how long we have been waiting for the tax cuts they have promised?"

"Tax cuts? No, really… There hasn't been any taxation since the days of the bombs… and anyway, have you looked around? The factory itself… it's been abandoned for decades, look there: the roof has collapsed, the *neige* is getting inside…"

"What do you mean by *neige*?"

Overwhelmed, Zam stepped back.

She imploringly turned towards the other three people, who in the meantime had never taken their eyes off the white cloth.

"You, you at least know what *neige* is, don't you? Do you remember the days of the bombs? Do you realize that this firm story is absolute nonsense…?"

One of the two women raised her head and turned towards Zam, slowly, her eyes like two lemons, an ominous smile crossing her face.

As they walked out of the factory, Typtri watched Zam, who ruffled her own hair until it cried with annoyance and anxiousness. He had seen the same absurd behaviour of those people in several other Remnants, most of all in the older ones. Never in the Runts, who found it much more natural to nurture a nearly Zen survival ability dripping with everyday hyperactivity.

"What matters is that we haven't found your fuse."

"I don't know if that is what matters most, Typtri. Anyway, let's walk around a bit more: we might find some more factories, a dump or a car wrecker…"

The deer was waiting for them near the gate, intently picking lichens from the trunk of a hibiscus tree.

When he saw them appearing, he immediately stopped, as though they had caught him trying his mother's clothes on. Then he hobbled in his disconnected movements and looked at them with an embarrassed look and a wide smile.

Stuck to the gate bars there was a white sheet of paper covered with an uncertain and childish calligraphy written in pen. When Zam got closer to read it, the deer's smile widened to an unlikely size and his chest started hiccupping with what Zam could have sworn was a laughing fit.

> Workers!
>
> The day has come when we will raise our heads against the arrogance of capital and the injustices of this bourgeois society. A fair and decent salary, better working conditions and the promise of an adequate future to our children—this is what they want to deny us, and this is what we will reconquer with a hard, stubborn fight. It will be no cakewalk, for none of us, let us acknowledge this, comrades, but through unity and resolve, the voices of every victim of exploitation will turn into a choir that will break through the wall of capitalistic oppression.
>
> No crackdown can stop us if we stay united!
>
> United Anticapitalist Workers' Movement

As soon as she finished reading, Zam heard a weird rustling coming from the ivy bush. As she approached it, she realized that someone was calling her.

"Young lady, beware of reading our flyer in such a brazen way. If the guard sees you, he could call the flics and accuse you of having posted it…"

"Excuse me? I'm afraid I don't understand, who are you?"

"Sssst! Not so loud, young lady! Can you see those CCTVs up there? They are directly connected to the control room, where dozens of flics are watching and recording all images! We are all under control! Control the controller, miss!"

Zam was getting dizzy. She turned towards the deer, who kept hiccupping in that clumsy way of his.

"Well, actually I'm not sure I get what you mean. Anyway, right, I'll try to be aware of this controller…"

A second voice emerged from the bush.

"Young lady, apparently you haven't opened your eyes yet in front of the barbarism that has taken hold of our country. Haven't you read about the last financial law, solely aiming at hitting the workers' wallets? Have you not heard the Minister's arrogant talk? Inside there our comrades risk losing their jobs! Dozens of them have already been made redundant!"

"Oh! You're talking about the guys inside there who are weaving those strange flags? But it's only three of them… and then you all keep talking about this minister… I'm sorry, but there are no ministers… there is not even a government… what's this financial law you're talking about? Has it something to do with finance? 'Cause, really, there is no finance either, there hasn't been for years… you could do a lot of things, you could build the society you like, I don't know how you manage to survive in this imaginary world…

"Young lady, it is you who live in an imaginary world! A world portrayed by the media, the capitalists' lies…! Young lady, what's your job? What professional category do you belong to?"

Zam walked away, upset as an Alice in a slightly gloomier Wonderland.

That deer kept laughing, but he must have felt that she was getting very annoyed, since he tried hiding his face to avoid being seen.

Typtri realized that Zam had had enough.

Third Installment

Mrs. Apricot eased her cup of tea on the dish carefully, so as not to produce any loud noise.

"After all, Mrs. Cerise, we live in incredibly unstable years."

"Unstable and dangerous, if I may express my opinion."

Mrs. Cerise threw a fleeting glance of cupidity to the last pastry left on the oval tray.

"Apparently General Saprofit is planning an extraordinarily powerful counter-attack."

"Do you think they will use new kinds of machines? A new model of mechanical man perhaps?"

Mrs. Apricot massaged the lace strip that wrapped her neck.

"I do not think so, my dear: mechanical men are so obsolete. My husband always says that those devices are only good for civilian purposes. As with your Zamedite."

"Well, we must admit that these mechanical men have been a gift of God to us. In these times of ill-omened scarcity, many a honest family would have been forced to make do without any reliable domestic service."

"My husband says that in the war fields many technological progresses are taking place. They say that since the cosmetic industry sold their recipe to the Army, our chances to win this war have become certainty."

"Even without any mechanical men?"

"And much better so, my dear!"

A giggle slipped from Mrs. Cerise's lips.

"Do not forget that mechanical men are men nonetheless, even if they have some additional optionals."

Zam entered the room silently with a new tray of pastries.

"Zamedite, my dear, could you also bring us some of those aniseed sugar candies our Mrs. Cerise is so fond of?"

"Oh, my goodness, you spoil me…!"

"Unfortunately, Madam, I'm afraid we're out of sugar candies and Mister Fleen behind the corner has no more of them in his store."

"God give me patience! All Fleen can do is complain. Go to the harbor, my dear, do me this favor. I am sure Mrs. Bienvenue's grocery store will not disappoint us."

Zam performed an imperceptible curtsey and vanished.

"*Your Zamedite is really a dear lass.*"

"I agree. These mechanical men are so tidy and pleasant. I really cannot understand why they are so badly mistreated in some popular circles."

"The mechanic's helper at the garage just outside your door: he is dreadful! That strange purple-eyed mutant gives me the creeps every time I walk past there."

"But, you know, those are Nature's mistakes. And to these mistakes we have tried to give a second chance. This has nothing to do with mechanical men, the product of masterly techniques. Mechanical men join man's intelligence with the perfection of machines."

This is what my husband always says, Mrs. Apricot. But I cannot resist these gorgeous pastries. Let me taste one more.

As she stepped down the stairs, Zam heard Mrs. Apricot and her friend's whimpering chat fade away.

She opened the heavy wooden door and went out, looking in the face of the sun while a soft breeze combed her wrists.

Lip was leaning against the crumbling garage wall, his lids almost hiding his weird eyes.

"Good morning, Zam. Are you doing errands for the old lady?"

"Hallo, Lip. Yes, I am. It's for their damned aniseed sugar candies. Normal people can't even find bread anymore, and they get distressed because aniseed sugar candy import has stopped. Now she wants me to go to Mrs Bienvenue's for the usual pantomime: 'Oh, Madam, are you really going to get some? We will certainly wait, Madam, don't worry. Sure, it is just a matter of days, Madam… Oh, yes, Madam, this war will soon come to its end…'"

"And will this war soon come to its end?"

Lip's eyes were glassier than usual, almost like liquid fog. His face was painfully absent, and it was made even more impalpable by the shadows that made it up.

They set off together, entering a street plastered with too-small stones.

"What would I know? No, I don't think so," Zam said. "What's sure is that none of these idiots will survive. I'm often sorry. For everybody, even for Mrs. Apricot. But to me it's like a sort of really painful evolution leap. They had a lot of chances, all of them, and they keep getting it wrong… I'm not sure about what we should do either, but it's that they always do exactly what they certainly shouldn't. Like extinction-bound lemmings, if you see what I mean."

"Yeah, well, I don't know what this world would look like if human beings became extinct. Quieter perhaps, and it would surely turn a lush green tinged with a lighter blue. It would also have the ferrous taste of blood, though, wouldn't it? The leap you're talking about will be a cynical meat chopper."

"I always forget how illogical feelings are. It's because of my mechanical side."

"That's your logical side, Zam. All that's mechanical in you is metal, but you aren't made of metal."

Lip still remembered the day when they brought Zam in for him to fix her. That memory was a white spot in his vision when he closed his eyes. He had never fixed up a human being or extracted metal cogs from anybody's shoulder blades. Zam had her eyes wide open and her hands clasped around her throat, her half-closed lips had held her breath for an exaggeratedly wrong lapse of time. Lip had replaced the broken parts, had softly detached her hands from her throat and had closed her lips. Then he had wrapped a blanket around her and had waited as she slowly woke up.

"Some days ago someone brought a robot to the garage. It was broken. Well, it should have been thrown away, actually. Those who designed it forgot a lot of things and the result is this precarious, bewildered thing—absolutely unserviceable."

At the garage where Lip worked, people often brought the remnants of broken prototypes that could be reused or simply dismembered for parts. It was a scarcely remunerative business: the firm threw prototypes away only when they had ascertained that they were totally useless. At the garage, spare parts were so abundant that generally machines and robots were simply heaped up in a lifeless pile in one corner.

"This time I managed to put my hands on it though. I persuaded my workmates I could turn it into a robot to keep the garage tidy, as a cleaner I mean. Anyway we haven't seen the boss for weeks, and with this crisis we aren't really flooded by commissions."

"...Oh, Lip! You have to introduce us!"

An unavoidable smile exploded on Lip's face.

"We're going there, Zam, you don't really think Mrs. Bienvenue can't wait to sell you her last aniseed sugar candies, do you?"

The stairs that led up to Lip's attic were castled defensively—a wooden spiral creaking at every step. But when you'd got over this necessary twilight, you'd open the door and almost collapsed before the light gust falling upon you from the huge skylights in the only large room of the flat. There wasn't a corner, a shelf, a workbench or a carpet that wasn't flooded with light. "It's because of my eyes," Lip used to say.

In the middle of the room there was a table and on the table, its feet firmly planted, there was the robot. It had just one eye, the other one being on the table, under repair. Its body was coarse, with too small, rather ridiculous feet, but despite it all its look was mighty and fierce. The big manufacturer's logo stood out on its chest, and its hands were complicated tongs with a network of metallic veins that bound the hands to the arms.

"Zam, this is Typtri."

Zam carefully wiped her hands on her apron, then softly caressed the robot's shiny shoulders.

Typtri looked at her with its one eye—it was a watery eye, full with an inevitable sadness that Zam did not know.

"It's not arrogant as the other ones: it looks nicer."

"It's the awareness of decline, I guess. Typtri never got to the end of the finishing programme; when they brought it to us, they said its eyes were its weak spot, too fragile and too expensive. Its irises are tangles of gears. Without them it would stumble everywhere and end up getting vulnerable and useless. New models are eyeless. They have sensors that perceive hindrances. They are much more basic, but are also substantially cheaper.

Zam's eyelids grew moist.

"I wonder what will remain of us weird beings when this war is over and all humans we know are dead."

Lip looked tired, but his eyes were now darker.

"We will be there, Zam. We have enough spare parts to try and outlive any surge of discouragement we may feel. We have our modified and mended bodies, your metallic nerves, my chemical eyes. Punks like us will have a world of ours to rebuild, the fullness of infinite days, the view of a time that has been turned upside down and sown with new seeds."

Lip's face was white, but Zam felt as if the shadows that crossed it were faster and lighter. In the many years she'd seen this room, it had never looked so bright. The dust covering Lip's tools had never seemed as delicate, nor sludge as sweet.

Zam reopened her eyes to see a detail of Lip's hand elegantly stretching in front of her eyes.

"Come, Zam, let's go to Mrs. Bienvenue's now. Mrs. Apricot cannot survive too long without her servants."

Ruggine might be the closest thing that SteamPunk Magazine has to a sister publication. Coming out of the Italian radical/squatter scene as SPM had its roots in the US, Ruggine *is a journal that sets out to use imagination to challenge the status quo. I was delighted to have a chance to interview two of their writers and editors, and beginning in this issue we will be running the occasional translation of stories from their pages.*

A COLLAR OF RUST

AN INTERVIEW WITH COLLANE DI RUGGINE

by Margaret Killjoy

SPM: *Can you introduce* Ruggine *to our readers? What does the name mean? What kind of stories do you run?*

REGINAZABO: *Ruggine* is a fanzine of radical fiction and illustrations. We publish short stories that try to dissect our tragic, pathetic world through irony and imagination, in the belief that metaphors can be sometimes stronger and more convincing than plain, objective, nonfiction essays.

We have a DIY approach to the whole process: our breeding ground is the Italian punk DIY community and the hacker scene, which are widely interconnected since hacker spaces in Italy are often inside squats and social centers. So not only do our authors and illustrators belong to this environment, but also the people who contributed to our graphic layout are at least acquainted with the idea of open-source. Our publications are released with no copyright, or at most under CC licenses, and we also try to use open-source instruments and DIY resources in the graphic layout and distribution: from the open-source software we use for layout (Scribus), to open fonts, to a distribution platform which allows us to find co-producers for our future publications. (*Produzioni dal basso*, a sort of self-managed Kickstarter ahead of its time.)

In the beginning we were planning to publish books, and our first publication was a short essay on the relationship between the humankind and technology in the light of J.G. Ballard's works. As we talked about our future projects, decay always was an object of our reflections, so we decided a good name for our DIY publishing house could be *Collane di Ruggine*, *collane* meaning "necklaces," but also "book series," and *ruggine* meaning "rust." So we had our rusty series, but no long text to publish in a book. Instead, we had a lot of short stories we had written and some others we had found in English on the web and wanted to translate. Switching to a fanzine was immediate, and we didn't have to think much to decide that it would have illustrations: we wanted something pleasant, that people would wish to pore through at first glance. The title was immediate as well: if the publisher was called "Rust Book Series," the magazine could only be, simply, "Rust."

PINCHE: *Ruggine* is a small DIY fanzine. We all belong to this weird community made of squatters, hackers, anti-psychiatric activists, punks, and similar creatures. One day we started to reflect on the fact that in this community there is a lot of non-fiction, that we create wonderful music, write amazing theater pieces, and build houses, but don't produce any fiction. We started to think that our community needed to recreate its imagination, which perhaps had been made too sterile by years of day-to-day struggles and frustrations. *Ruggine*'s authors are very different people, but somehow they share a passion for "imaginary literature" (we like to use this expression, by Italo Calvino, since it adds a different scope to the idea of science fiction) and love to plunge reality into a richer dimension. *Ruggine* never had a determined "editorial line": we choose the stories we publish by inhaling them, looking for that imaginary atmosphere that goes deep to the heart of the matter even when the plot is about apparently far-away situations. We look for a way of writing that is somehow archetypical, that talks to people in a way that is ironically more direct than a thousand news articles could be.

Rust is one of these archetypes: it is the result of a something that used to be a certain material but is now changing into something else, it is the past turned into future in a painful, beautiful way. We imagine the exploded world of this non-future of ours as a rusty world covered with vegetation.

SPM: *What is the steampunk scene like in Italy? What is it like interacting with that scene as radicals? What about the other way around: what kind of response do you get with your fiction from the radical scene?*

REGINAZABO: I don't know of any "steampunk scene" in Italy: there may be some cos-players, there are a couple of websites, but I've never heard about any steampunk public events. Frankly, the only steampunk parties I've been to had been organized by people connected to Collane di Ruggine and therefore to the punk/squatter/hacker scene. I've tried to get in touch with some steampunk authors, but I found them a bit disappointing, perhaps because I hadn't understood yet that steampunk is often definitely not punk, not only in this country. In Italy most steampunk literature is actually genre literature with a Victorian twist, so you'll have some Victorian fantasy novels and some Victorian thrillers, but nothing particularly impressive, as far as I know. I find that steampunk has a lot more potentialities, I've tried to explain why in my introduction to the Italian translation of *A Steampunk's Guide to Apocalypse*, but in short it offers us the possibility to tell our history from a different, radical point of view, and I'd like to see the literary steampunk scene do a lot more of this.

On the other side, *A Steampunk's Guide to Apocalypse* (which wasn't published by Collane di Ruggine, but came out together with the first issue of *Ruggine*, containing a story and illustrations from *SPM* #1) was a great success in Italy: many people here have learnt about steampunk from this book, and I'm happy that their approach to this subculture was radicalized straight away, mixing from the beginning the steampunk aesthetics with a punk DIY approach. In the radical scene, we've had a good response: many people send us their stories and invite us to present our project in social centers, but I think this has less to do with steampunk (which is very appreciated nevertheless) than with our general approach to radical fiction.

PINCHE: When we read your vision of steampunk in the pages of the *SteamPunk Magazine*, we thought that this was what we actually were: strange creatures that are more punk than steam, and are firm in their decision to reassert the tight bond between both things. But unfortunately we haven't found this spirit virtually anywhere in Italian realities. I can think of a couple of Italian mad scientists who toil in their dark cellars without knowing that they are ideal steampunks. The few things that exist are often soaked in (the worst) pop culture, narcissism, and aestheticism, gaudy and not very interesting.

The presentations and reading we do with *Ruggine* nearly always take place in radical venues, and we have often had positive responses even if we are not very well known. I think there is some resistance to a vision that has so little materialism in it, but there is also a lot of curiosity and a certain sense of liberation in the possibility of using expressive means that go beyond the conventions of the "perfect modern squatter." *[Editor's note: I believe Pinche is referring to the "perfect modern squatter" as what is accepted unquestioned by many in Italy to be the most legitimate form of radical activism.]*

In the last few years we have regularly participated in the Italian Hackmeeting, where I feel we are perceived as a vibrant part of the community because of the common way in which we look at technique and at self-management, and because of our smiling apocalyptic vision. My aim is to have *Ruggine* read by the crustiest punk of the suburbia, instead of seeing it circulating in sophisticated readers' clubs, so I guess I can be pretty satisfied for the time being.

A superb project for those who have a habit of keeping fabric scraps for a rainy day, or for those who scavenge markets and antique stores looking for textile treasures, this project is suitable for anyone who has a little patience and a desire to make their very own piece of wearable art.

Introduction

Often found on adventurers and inventors alike, the wrist cuff makes a delightful addition to any wardrobe. Featuring the foundation of what you will need to make a basic wrist cuff with a few embellishments, this accessory is perfect for both gentlemen and ladies, with or without the lace.

After learning the basics, feel free to dress up your cuff (or dress it down) as much as you'd like. The beauty of cuffs is that they are amazingly versatile, so once you know the basics of making one, the sky's the limit!

Measure

To begin, measure the circumference of your wrist and add an extra 1/4" to that measurement. This will be the *length* of your cuff. (You can add more than 1/4" if you'd like your cuff to fit more loosely.)

Cut the pieces of your fabric or felt to the length of that measurement. Then cut each piece so they are both 2 1/2" wide. Cut your brocade (or alternative) to the same length, cutting the lace to match.

SEW YOURSELF A CUFF

by E.M. Johnson

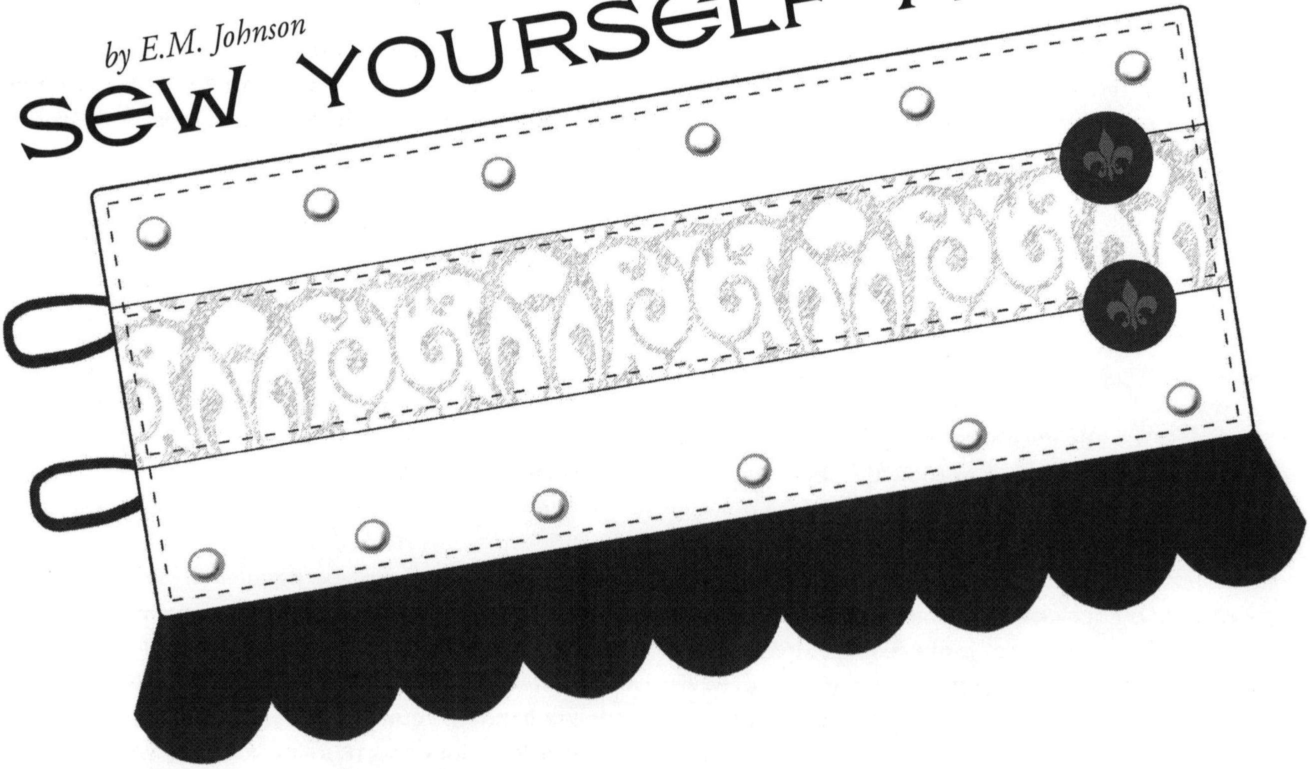

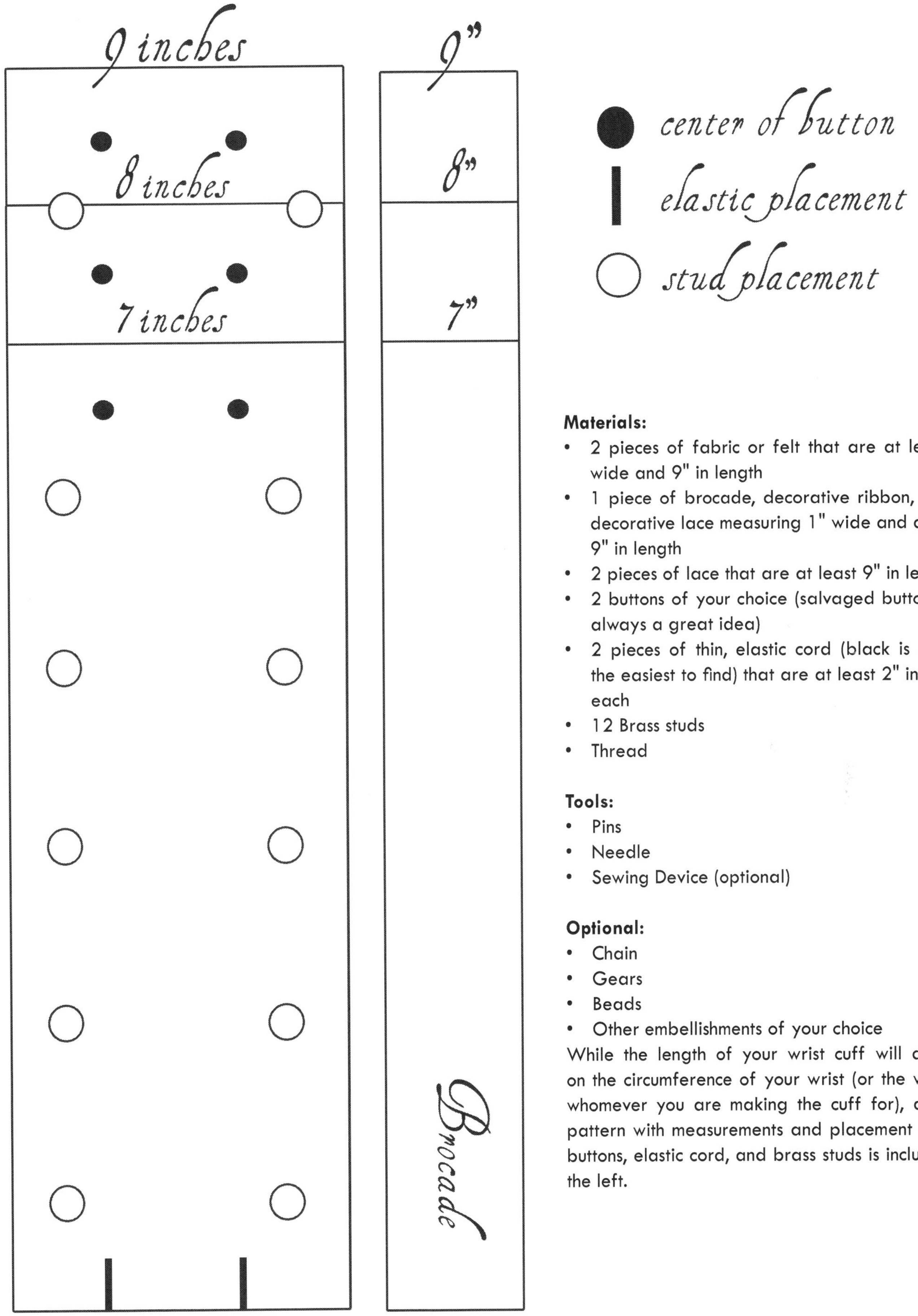

Materials:

- 2 pieces of fabric or felt that are at least 3" wide and 9" in length
- 1 piece of brocade, decorative ribbon, or flat decorative lace measuring 1" wide and at least 9" in length
- 2 pieces of lace that are at least 9" in length
- 2 buttons of your choice (salvaged buttons are always a great idea)
- 2 pieces of thin, elastic cord (black is usually the easiest to find) that are at least 2" in length each
- 12 Brass studs
- Thread

Tools:

- Pins
- Needle
- Sewing Device (optional)

Optional:

- Chain
- Gears
- Beads
- Other embellishments of your choice

While the length of your wrist cuff will depend on the circumference of your wrist (or the wrist of whomever you are making the cuff for), a basic pattern with measurements and placement for the buttons, elastic cord, and brass studs is included to the left.

Embellish - Part 1

Take the piece of fabric you want to use as the top of your cuff and pin the brocade/ribbon so that it is running lengthwise down the center of your cuff.

Using either a sewing machine or a needle and thread, stitch your brocade/ribbon so that it is securely adhered to the fabric.

Closures - Part 1

Lay the second piece of fabric face down, so that the "good side" (the side of the fabric that you want to show on the outside of your cuff) is facing your work surface.

To measure the length needed for your elastic cords, take one of your buttons and make a small loop with your elastic that is just big enough for your buttons to fit through.

Once you have determined the size needed for your elastic cords, pin the elastic cords to one of the shorter sides of your fabric with approximately an inch of space between them. After they are properly secured, stitch them in place.

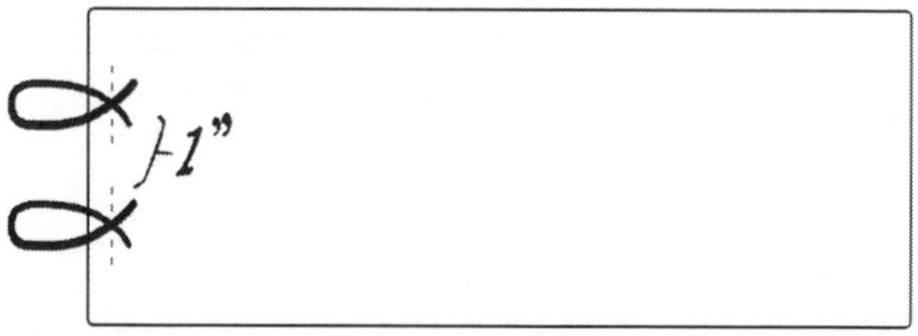

Embellish - Part 2

Skip this step if you don't wish to add lace to your cuff.

Keeping your fabric so that the "good" side is still facing the worktable, take your piece of lace and place it at the bottom portion of the fabric so that the straight, non-ruffled portion of the lace is overlapping the edge of your fabric by 1/2".

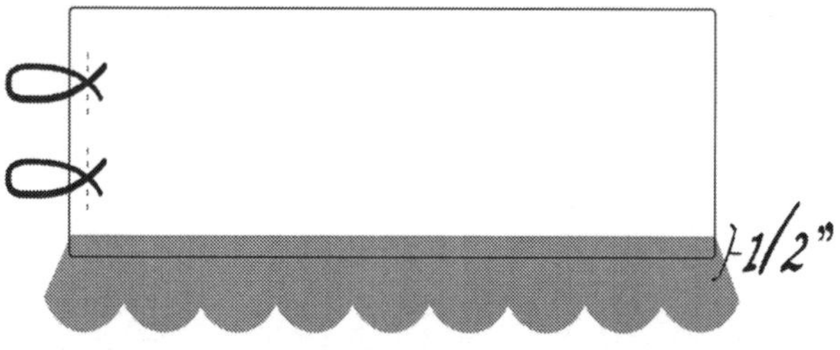

Construction

Now place your other piece of fabric (embellished side up) directly on top. Pin both pieces of fabric together, paying special attention to make sure the lace is securely pinned in place between them.

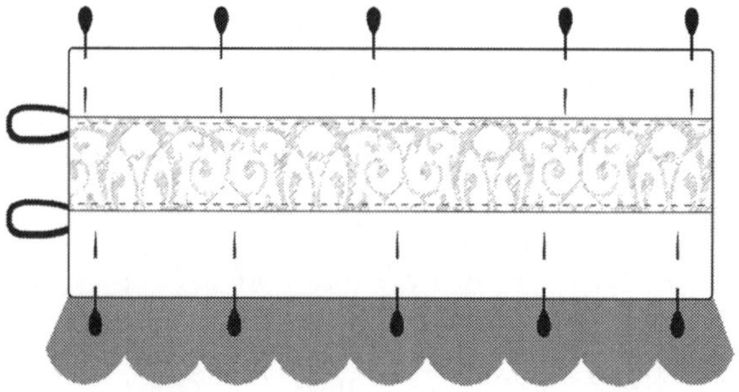

Leaving an outer hem of 1/4", sew around all four edges of your cuff.

Note that the edges of the cuff will fray over time, adding to the antiquated look of the cuff. However, if you don't want to wait for Father Time to work his magic on your cuff, you can fray it yourself by ruffling the edges of fabric until they begin to fray.

Embellish - Part 3

Add brass studs approximately 1" apart along the long sides the cuff (inside the hem) following the manufacturer directions.

You can space studs closer or further apart depending on the size of your cuff, and how many studs you would like to have for embellishment.

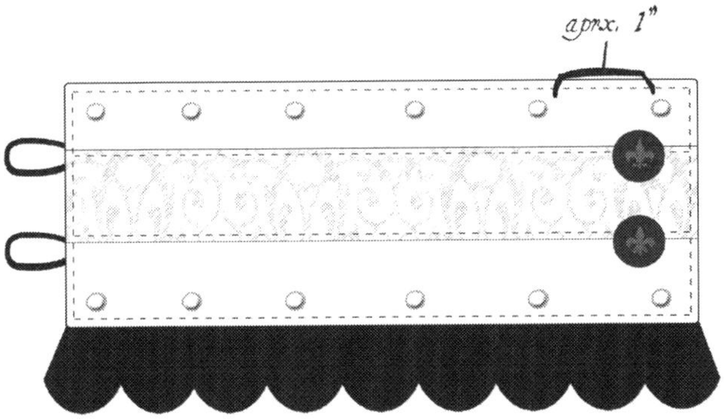

Closures - Part 2

On the opposite side of you cuff, line up your buttons so they match the location of your elastic cords. Move the buttons away from the edge of your cuff until the center of each button is approximately 1" away from the edge.

Congratulations; you now have a wrist cuff to call your very own. But why keep it all to yourself? Wrist cuffs make wonderful gifts.

Variations on the Theme

Now that you know how to make a basic wrist cuff, it's time to personalize it with other items and embellishments if you so desire.

- Sew chain and gears along the cuff
- Add lace to both sides of the cuff
- Swap the lace for chiffon, or layer chiffon under the lace
- Consider adding beadwork or decorative glass
- Sew charms into the lace

References

ashleyannphotography.com/blog/2010/11/23/diy-fabric-cuff/

www.createmixedmedia.com/wp-content/uploads/2010/12/CuffBraceletProject.pdf

A Dieselpunk Primer

by Larry Amyett, Jr

Steampunk, which is the primary subject of this magazine, is actually one of several subcultures of genre-punk. While most readers certainly know about cyberpunk, there exists a lesser-known genre called dieselpunk.

Over time, the consensus of the dieselpunk community has defined dieselpunk as essentially consisting of three components: having a characteristic called Decodence, being contemporary in origin, and containing the element of punk. When combined, these three components create something that is altogether new and, as the saying goes, is greater than the sum of its parts.

One mistake that some individuals make is to reduce the "diesel" in dieselpunk as simply being a reference to the dominance of the internal combustion engine. However, that would be the same mistake as saying that the "steam" in steampunk makes the genre nothing more than the use of steam technology. Just as steampunk is something more than the steam engine, dieselpunk is more than the internal combustion engine. The best explanation of the word diesel in dieselpunk is through the understanding of the concept of Decodence.

The term Decodence is a portmanteau from the words deco and decadence. Originally coined by dieselpunk pioneer Nick Ottens at the website The Gatehouse (www.ottens.co.uk/gatehouse), Decodence is the essence that surrounds everything of the time of the 1920s through the 1940s, which is a period that dieselpunks call the Diesel Era. Decodence includes, but isn't limited to, gangsters, jazz, big band and swing music, the Great Depression, G-men, art deco, noir, prohibition, fedoras, flappers, zoot suits, hardboiled detectives, tommy guns, Nazis, zeppelins, World War II, and all of the other historical and cultural elements that distinguished that era.

For dieselpunks, Decodence is vital. A review of the various dieselpunk web sites and blogs will reveal what may appear to the visitor to be a near obsession with anything and everything of the Diesel Era. Many dieselpunks exhibit an insatiable curiosity about even the most minute detail from the 20s–40s. Some might argue that this is simply because the Diesel Era left a greater wealth of audio-visual material in comparison to previous periods. Certainly, there is some truth to this because the technological advances in the Diesel Era allowed for the survival of material that wasn't previously possible. However, for the dieselpunk it's not the fact that it exists but is instead a fixation that dieselpunks have for Decodence.

If we simply stopped here, we wouldn't have dieselpunk but simply an infatuation with history. This takes us to the next major component of dieselpunk, which is that it's contemporary in origin. Even though Decodence is very important for dieselpunk there is an equally strong emphasis on incorporating it into new creations. Decodence is to the dieselpunk what clay is to the potter in that it provides the source material in the creation of something that's new and original.

However, simply combining Decodence with contemporary, while both important to the genre, doesn't result in dieselpunk. Without something more then all we would have would be a hobby of historical reenactors participating in a form of living history. Dieselpunk needs one more component. It needs an element that adds power and vitality to it. It needs punk.

At this point, it might be a good idea to explore the historical origin and various meanings of punk. According to authors, David K. Barnhart and Allan A. Metcalf, the roots of the word date back to the Native American word "pungough." Some tribes used pungough as the name for a powder that they made from burning corncobs that they then mixed with their meals while others used it to describe ashes. Over time, the word was Anglicized into "punk" where the early meaning ranged from burning sticks to cigarettes. Author and humorist George Ade once used it to describe something as being worthless. By the Diesel Era, the meaning of punk had evolved into what Barnhart and Metcalf called a "small time hoodlum." Punk music and subculture appeared in the 1970s, reveling in its image of being outside the mainstream.

This etymology of the word punk provides us with the understanding that punk is something that outside the mainstream to the point of being considered worthless by "respectable" society. It's something that's non-conformist and not part of the establishment. To be punk is to be an outsider.

By adding punk to Decodence and contemporary, we now have the makings of a unique genre that we can call dieselpunk. This is because punk allows for something more than studying history that Decodence alone provides. In addition, with punk there is now the potential for something more than reenactment of a bygone era. Because of punk, a potential now exists for alternative history, fantasy, and speculative fiction. Punk is the component that allows for the exploration of "what if?" Dieselpunks love to discuss scenarios such as what if the Hindenburg had not crashed and ended the age of airship travel or what if the events of World War II had unfolded differently.

Now we have the necessary components for a completely original and unique genre-punk apart from either steampunk or cyberpunk. Dieselpunk uses the history and culture of the 1920s through 1940s as source material and filters it through modern sensibilities while combining it with the power and vitality of punk. With this understanding, we can now explore how these components come together.

Dieselpunk manifests itself in different forms. The styles or flavors of dieselpunk range from the positive to the dark and even to the nightmarish. In dieselpunk, one can explore the optimism that prevailed during so much of the early Diesel Era or travel down into the deepest horrors imagined, which too were part of the reality of the time. By its nature, dieselpunk has a potential for creative expression and serious exploration of all aspects of human existence.

There's also a strong characteristic of individuality in how each dieselpunk views the genre. While most dieselpunks generally enjoy all variations, dieselpunks vary in their preferences. For example, one dieselpunk may prefer the style known as Hopeful Ottensian, which recalls the exuberance and belief in unlimited human progress that was so prevalent in the 1920s. However, another dieselpunk may prefer the style of Dark Ottensian, which focuses on the dark underbelly of the Diesel Era, filled with gangsters, dominating capitalists, economic hardship, and war. Either of these styles, as well as the other variants, are equally valid manifestations of the genre. Therefore, while I've provided a consensus of dieselpunk in this article, ultimately one must ask the individual, "what is dieselpunk to you?"

As is often quoted, George Santayana wrote, "Those who cannot remember the past are condemned to repeat it." Dieselpunk aids us in learning from the past and better equips us to address the challenges of today. By examining the disastrous failure of prohibition, for example, dieselpunk helps us to have a better understanding of the implications of the modern "War on Drugs." Another example would be that the dieselpunk interest in the Great Depression and the Diesel Era response to it, such as the New Deal, provides us with resources by which to examine how best we should respond to our current economic crisis.

Dieselpunk, which is a young genre (the term dieselpunk was coined in 2001 by game designer Lewis Pollak), is experiencing an explosive growth in popularity. Online dieselpunk forums are seeing continual increase in membership and dieselpunk blogs are gaining more and more readers. Various print and online magazines, such as this one, are incorporating articles concerning the genre. The genre of dieselpunk is very international and dieselpunks in various nations are currently in the process of organizing face-to-face meet-ups.

Dieselpunk artists include Stefan Prohaczka, Alexey Lipatov and Keith Thompson just to name a few. Bands such as Big Bad Voodoo Daddy, Tape Five, Caravan Palace as well as musicians such as Wolfgang Parker are all considered dieselpunk. In the cinema, we find dieselpunk movies such as *Sky Captain and the World of Tomorrow*, *The Shadow*, The Indiana Jones series, *Dark City*, *Captain America: The First Avenger*, and *Brazil*. Examples of dieselpunk literature include *Hard Magic* by Larry Correia and *Fistful of Reefer* by David Mark Brown.

The future of dieselpunk is exciting and positive. If you're interested in learning more about the genre there are several online sites one can visit:

"Dieselpunk" (my blog) DIESELPUNK44.BLOGSPOT.COM

"Dieselpunks.Org" (one of the largest dieselpunk Forums, moderated by influential dieselpunk Tome Wilson) WWW.DIESELPUNKS.ORG

"Flying Fortress" (one of the first dieselpunk blogs) FLYINGFORTRESS.WORDPRESS.COM

"Smoking Lounge" (forum moderated by dieselpunk pioneer Nick Ottens) WWW.OTTENS.CO.UK/LOUNGE

"Dieselpunk" (blog by Lord K, an influential dieselpunk) DIESELPUNKS.BLOGSPOT.COM

The Scouts of the Pyre

By David Z. Morris
Illustration by Scary Boots

With profuse apologies to the body and spirit of Joseph A. Altscheler

Part 1 of 3

Captain William Sherburne told Lieutenant Harry McGee that the invaders were coming, and there was a stir among the ranks of the defenders. The riflemen, disciplined and weary and bored, said nothing. Most did not even turn downrange when the whispers reached them, only continuing to settle their blades or check their pistols or scribble a few words into their daybooks. By contrast, the new recruits craned their necks towards the line of woods and, despite the commands of their various bullheaded sergeants, exchanged apprehensive whispers.

Harry walked with Captain Sherburne down the road just in front of the barn—a barn that was now a weapons cache, and a target. The Captain scratched at a sideburn with one long finger, gazing not toward the advancing enemy but at the ground. The Captain's blue uniform was worn, if not quite ragged, but Harry hoped he still looked half as commanding in his own."If only we had fire," Sherburne lamented in a low voice, "we could be certain."

"Yes," replied Harry, "The loss of that cannon will not be remedied soon. But do you anticipate real trouble? It's only a small magazine."

"Small, but… important. I hope to God the Confederates don't know what is here, and we can brush them off. If they take it…" Sherburne trailed off hazily, lifting his gaze to the sky. Then, seeming to shake off his dark malaise, he looked at Harry with sudden sharpness. "The rifles and carbines and blades will have to do, however ugly things may go."

Harry nodded firmly, then followed Sherburne as he continued walking.

All was silent, though the scout had reported the enemy close. These were the worst moments, Harry thought—knowing what lay ahead, and being impotent to act. More men were known to break in the moments before such an encounter than during it. Thankfully, the captain made a good show, seeming utterly unperturbed—almost preoccupied, as if the fight coming to them was a mere nuisance. Harry, by contrast, felt himself a ball of nerves, barely resisting the urge to turn his own head expectantly towards the woods. He only succeeded because he knew the men were better eyes than his own could be.

He was not wrong. Only a few moments later their muttering crescendoed, and one fresh recruit even raised his carbine to his shoulder. A sharp-eyed sergeant immediately reached out his swagger stick and tipped up the barrel. "What would you do," he asked the trooper with a sardonic grimace, "*Scare* them away?" A sick and brutal laugh went up from more experienced men along the line, and the recruit lowered his weapon with a mix of sheepishness and terror.

The terror was understandable. For now Harry turned to see them, just barely visible, scattered among the trees—low, grey shapes, moving quietly forward. In a few more moments, they would be within the range of the experienced marksmen along the flanks, but for now there were no reports. As the Union whispers died in the moment's shift from nervousness to anticipation, there was only one sound, creeping through the trees towards the waiting soldiers:

Moaning.

Harry had heard it before, as had many of the men, but this was no aid. It only meant one knew what came after, the relentless waves, the slow exhaustion, the inevitable blood and misery.

Moaning.

It was the sound of a loss beyond loss. The lament of one who did not understand his own sadness, who had no desire but to know what he had once wanted. At this distance, Harry realized for the first time, the few lonely wisps of inarticulate despair were poetic, each telling the tale of a life lost—and more than lost.

He did not think the empathy worth the nightmares that had bought it.

"READY!"

The staff sergeant's cry rang out. In only seconds the range would be closed, and the storm would begin.

"AIM!"

In that last scrap of silence, the moans from beyond the treeline lost their individual character. The enemy was no longer a collection of sad tales, but a mindless chorus that sang threat and doom. Harry finally walked up to the line, and saw what he had seen too many times before.

Shambling, mindless. By the dozens and dozens. Corpses, walking like men.

"FIRE!"

The riflemen fired first, while the sergeants in the middle of the line held the new recruits and their carbines at bay. Before the smoke fully closed in, Luitenant McGee was able to see far too clearly what happened when the marskmen found their targets. Grey, desiccated heads cracked or exploded, one here, one there, sending sprays of white bone and grey matter so forceful as to sometimes make a nearby invader stumble. But not once did the enemy stop to aid a fallen comrade, or hesitate in any other way. Sometimes even decapitated bodies would take a further wobbling step or two before tumbling pathetically into the dark, wet loam.

There was neither fear nor hate in the oncoming eyes. Even more than their pinched and hairless faces, their ragged clothes, their awkward gaits, it was this flatness that made it hard to remember they had once been men. An errant bullet would rip through a shoulder or calf, and the creature would continue without yell or grimace or pause.

As these grim pantomimes played out, Harry felt a shudder pass through the younger ranks, still hanging fire as the horrors slowly shambled into carbine range. He glanced quickly at Sherburne, who answered with a half-absent nod. Harry moved to stand just behind the men, and spoke in a voice firm and confident beyond his own inner sentiment, clear and loud enough to withstand the rising cadence of rifle cracks.

"They are terrible, and they are pathetic. You might think them worth your mercy, as sad as they are. But as you face them, remember what we are really fighting—not these things, but the forces that made them, the forces that wield them like a hammer and torch to burn down our Union. Remember that—they are mere things we destroy today, not men. The men once housed in these husks were killed long ago, by those very fiends that now drive their soulless shadows forward! Remember, whatever race the remains that shamble towards us, you are not destroying these men. You are avenging them!"

Suddenly embarrassed by the mounting of his own manufactured passion, Harry stepped back. A momentary silence hung among the young men. The sergeant then stepped forward and finally gave those boys the command to fire.

The destruction and gore of the next minutes were unimaginable. The unskilled recruits had little hope, even at closer range, of reliably landing the effective shot to the head. So they had been given carbines, guns that could produce a greater volume of fire than a rifle, but were much less predictable. Balls sailed through the bodies of the risen nearly at random—through shoulders which burst asunder, ribcages which collapsed backwards, hearts that threw gouts of grey sludge, and thighs that cracked and buckled. But until their heads were destroyed or detached, the things continued to advance.

The carbines began firing at something like thirty yards. This gave very little time to whittle down the enemy before he came upon the line, but the first volley sufficiently pushed their mass down and back. There was a brief pause for reloading, and a few muffled celebrations when the men again raised their heads. It seemed that the enemy had gone, or been annihilated. But it was only an illusion of the smoke now being blown downwind into the forest, and the incipient cries of celebration died stillborn when pallid arms and lipless mouths again began to swirl into slow attack.

It continued for some time. A wave would be downed, and the troops would barely reload in time to repulse the next. Harry thanked God that they were, after all, defending a weapons store, and had plenty of ammunition—but he knew that the men's stamina was not so plentiful. Soon, inevitably, the rate of fire slowed. The more experienced riflemen quickly pulled towards the center of the line to reinforce it, and Harry and even Captain Sherburne drew arms.

"They always *seem* infinite," Sherburne grimly observed to Harry, between volleys. "This time, I fear that impression may prove fact."

Harry sighted between two milky, vacant eyes and pulled the trigger. "Sir, is there something you wish to tell me?" He could make little sense of Sherburne's grim proclamations, but was glad the men were too preoccupied now to note their commander's mood.

"No, no," replied Sherburne, doggedly shaking his head. He had lowered his own pistol, with its ornate brass scrollwork and eccentric reloading mechanism, and gazed into the distance with a new intensity. "Just know that what we are fighting for today is important."

"I know that about every day that I fight, sir," Henry replied.

Once again raising his sidearm, Sherburne turned to regard him. "You are quick with fine words, McGee. Today, pray that you may be simply quick."

At that moment, there was a burst of panicked shouting from the line. The marching corpses had finally outpaced the carbiner's ability to reload, and were beginning to reach the men. The beasts were in horrific shape, missing arms and jaws and with guts dragging behind them like fallen banners. One of the attackers drug himself—*itself*, Harry revised the thought—along the ground, both his legs having been lost to wild shot.

But whatever their shape, these were still threats, in more ways than one. The riflemen who had reinforced the center of the line were now ferociously and valiantly bayonetting the creatures, aiming high and thrusting hard through eye sockets and soft palates. But the recruits, many of them facing the dead for the first time, were crumbling. One boy had stumbled wildly backwards from the charge, and now lay on his side behind the line, clutching a silver locket and repeating the name "Anne" again and again. Another, braver but less wise, had pulled out a long knife and taken the battle to the enemy—only to have no-longer-human teeth sunk deep into the muscles of his neck.

Harry took all of this in, watching in horror as the recruit—a mere boy—fell with a surprised groan, his bared white teeth splashed red. Still smarting from the Captain's rebuke, his mind was a volatile mix of vengeance and pride. He drew his saber and charged to the line—if Sherburne disdained his schooling in words, perhaps his schooling in blades would garner more respect.

But no sooner had he sunk the fat sword deep into the skull of the boy's killer, than he heard Sherburne bellowing from behind. "McGee! Back!" Harry couldn't resist laying out two more walkers before extricating himself, and said a small prayer that the remaining men could handle themselves.

"Whatever you do, remain calm," Sherburne commanded cryptically as Harry, already heaving like an ox, returned to him. Then the Captain extended a finger back past the frantic battle line, into the woods. "Look," he said, suddenly hoarse, quiet, and tremulous.

Harry turned, straining to tune out the rising screams and sickening noise of the battle as he followed the Captain's hand. The smoke was now so thick he could barely see the first line of trees, less than forty yards away.

"Sir, I don't know what you–"

And in that moment, he saw them. Not through the smoke, but *above* it—thick black tendrils, one after another, flowering out of the haze and disappearing back into it. Harry thought of the anemone he had spotted once in a tidepool in Nantucket, its slow, hypnotic movement in the waves, its predatory intent. These were little but outlines, menacing rumors of something else that remained hidden—something enormous.

"They have a goat," the Captain intoned, his throat clenching in barely-mastered fear.

Harry was dumbfounded, terrified, but also incredulous. "Why… why would the rebels commit one of the Young *here?*" he demanded, turning from the terrifying sight to Captain Sherburne.

Sherburne ripped his eyes away from the treeline, and they burrowed into Harry's. "McGee, you must go now. You must deliver word to General Grant of what has happened. Better yet, to the President." He gripped Harry's arms and gave them a firm shake. "There will be patrols, and you alone can evade them. Word *must* get back."

Harry was confused, and for all his confidence and anger, his fear was mounting. "What *has* happened here?" He asked, flustered. "Or what *will*? And if you are so sure of a loss, why sacrifice these men?" Harry choked this last into a whisper, only too aware of the battle still in the balance just feet away.

"I have not told you the full truth, my boy," Sherburne replied, his head tilting and his brow furrowing. "There is something in that barn. You could call it ammunition. But the… gun… that will fire it, is many hundreds of miles away. In Atlanta. From there, it could deliver a devastating blow to the Union. I and the men must stay here to slow its capture, for every second counts. Perhaps we can fight it to a siege…" Sherburne again seemed on the verge of sinking into morose reverie, then snapped back into focus. "But regardless, you must go, and go *now!*"

The force of the declaration was unmistakeable—no longer a friend offering information, Sherburne was now an officer delivering an order. Lieutenant Harry McGee, a proven warrior and true son of the Union, gave Sherburne a sharp, soldierly nod. Finding no words on his tongue or triumph in his heart, he fled the field of battle, and did not look back.

Not even when the screams began.

Harry found his own horse from among those tied up on the far side of the barn. He quickly mounted, and was soon riding through the village nearby. He feared no patrols so close, but knew he would soon have to go more carefully.

As he gained speed along the road, a man's angry shout came from one of the windows along the main road. "You bring them here, and now leave us to the horror?" Harry wished he had time to stop and tell the man his purpose, for he hated to be thought a deserter or coward, even by a stranger.

But at the far edge of the village, he saw something along the side of the road, near the very last house before the road made its way across a creek. White legs projected from a bush, prone. A spine-chilling stroke of crimson stretched to the middle of the road. For all his hurry, Harry could not help but dismount and move to help.

He could now see it was a boy, redheaded and fair, feebly dragging himself into cover. "Son, what has happened to you? Let me help." Harry reached out, gently turning the boy over onto his back—and then recoiling in horror and despair.

He had expected the blood, even the wound, ragged with muscle and skin and shining white globules of fat. But he had not expected the boy's eyes, shot with blue and fog, or his skin, its purple tint just going to grey.

The eyes moved fitfully, as the boy's hands gripped the empty air like claws. He took a ragged breath. "They're… all… around," the boy said. "Stay on… the creek."

Harry nodded, and could not resist placing a reassuring hand on the boy's pallid cheek—however great the risk. "Thank you, son."

And then he drew his pistol, and placed it against the boy's forehead, where it was just brushed by the bright, straight red hair. Not yet gone, the boy closed his eyes, as if ready to receive a benediction. A roar came then from the direction of the storehouse, a hundred mixed voices of animal and human rage, so loud that it almost drowned out the shot.

Harry wiped the barrel against the grass and took the horse's lead, nervously glancing back the way he'd come. He could not risk the noise of riding, but he would need the speed later. He scanned the surroundings unceasingly as he made for the creek, and silently said a blessing for the poor boy's soul. There was no other movement, nor sign of inhabitants.

On reaching the creek, Harry remained vigilant. The Confederate's unholy forces could not move quickly, but they were damnably close to silent, particularly in small groups, their moans and shuffles dispersing amidst the wind. Now, in particular, the receding, horrific noise of the battle had so much of the character of death that he had to work to maintain his focus.

He walked down the edge of the creek, picking his way through and around low cover and trees. He hoped that there had been at least some sense to the Captain's plan, hoped even more that the Captain had been wrong about the enemy's strength. But no, he was no naïf—he had seen the limbs of the enormous abomination, and knew that there was little mere men could do to hold arms against it. If they'd had the fire cannon, or even a single detachment of Special Cavalry, there might have been a chance. But as it was, Harry found his thoughts again on the poor boy, his single corpse now turned to dozens… and one with Captain Sherburne's bloody face.

Twice he was snapped from his grim reverie by the appearance of small wandering bands. Each time he froze against a tree and gently stroked his horse's muzzle to soothe it. Each time the group of five or six dead passed by, with excruciating slowness, but without raising an alarm. The golems, after all, were nearly insensate.

After almost an hour of this, Harry reckoned he had made it beyond any reasonable perimeter of the battlefield. He craned his neck beyond the treeline, where everything looked clear. He breathed a mild sigh of relief, led his horse up out of the creek and mounted, ready to make eastward for General Grant's camp. Maybe he could muster reinforcements in time to save his unit—or at least to pursue the attackers as they took their booty to Atlanta. Harry knew it was wrong of him, but he did feel some hint of pride mixed with his despair as he–

"Lieutenant McGee, a moment of your time!"

The cry rang out from the left, and Harry pulled up sharply. He turned, and was shocked to see a figure standing where he had looked just moments before. It was tall and slender, with a strange type of long black coat. Its hair was ragged and unkempt—and stark white. But most notably, the long sleeves of the coat clattered gently with the fetishes suspended along its underside—vials, and needles, and charms.

Harry gripped the flanks of his horse tightly between his legs. His entire body tensed. His pride, as so often before, had been premature.

"Commander," Harry replied grimly, "where are your troops?"

"That hardly seems important," came the reply. The man pulled himself up straighter. "And that's *witch*-commander, if you please. Perhaps it is silly, but we southerners are expected to abide by formality." He gave a slight, flourishing bow, his lips twisting into a smirk as his knees dipped just so.

Harry did not wish to leave any leeway for the foe before him, but knew also that an attack must be coming from behind. He in the end chose to conceal his fear under an unwavering gaze. "To call yourself a witch is an insult to the good people of Salem. But then, I suppose you know that." In an uncharacteristic and ill-timed show of pique, Harry punctuated his comment by emphatically spitting on the ground.

"Sir, you insult me twice!" The man spoke in a sardonic, lilting tone all out of tune with his grim appearance. "And what of it, if we have given the arts of the North a somewhat more southerly flair? Are we not still of the same family?"

The tension of awaiting the inevitable attack was quickly wearing on Henry. The hairs on the back of his neck stood on end, and he inched his hand towards the butt of his pistol. "Family? You bandy words with me, betrayer! And the South has no arts but that were stolen from its slaves."

The witch-commander took a few slight sideward steps, casual and slow. His trinkets rang, lonely and high, and again he smiled.

"Betrayer… slave… you should be more polite when discussing the current balance of power, and our peculiar institution. As for impure arts, I think you may protest too much—it is known you yourself are not what you once were."

McGee seethed, the hand nearing his gun nearly vibrating with rage and desire. "That is slander, a Confederate smear! I am sorry, Stephen, but I must put an end to it." And with that, he went for his sidearm.

In that instant a hail of dirt and stone spat upward in front of him, and his horse buckled, rendering that precious, quick shot less than a prayer. Harry looked down, and saw the soil convulsing, seething—and then he saw the teeth, and then the limbs of his beloved horse reduced to splintered scraps.

"You ask after my troops, brother, and I show them to you," quipped the enemy officer, shouting over the keening, frantic death cries of Harry's proud mount.

"Stephen!" McGee shouted in a rage as he scrambled to extricate himself from the saddle. Mourning would come later. "Why do we play this deadly game? What would mother and father have thought of your evil?"

Stephen McGee strode forward, declaiming with a sudden seriousness over the horse's bleating. "You think this is a game, but it is not. The Southern way of life must be defended. Lincoln and his lot want to rob us of our livelihoods and of our culture. And you speak of evil? Father would have stood with me right now, even against you."

The voudoun—for that was how Harry thought of his estranged brother's divergent arts—now stood mere steps from Harry, who had wrenched one foot up into the dip of his saddle, preparing to leap clear of the terrifying teeth that swirled beneath him. Harry, his concentration thus occupied, did not see as the Confederate raised one hand menacingly above his head and furrowed his strange brow. Did not see as smoke began to spew from his sleeves and billow from his loose collar.

"You think this is a game," Stephen intoned, "but it isn't."

Harry leaped, becoming aware only in that instant of the sphere of black death flying towards him. It caught his right boot as it hung in midair, and the world seemed to come to an entire standstill as the foot was sheared into nothingness just below the ankle. Even worse, the impact was enough to cut Harry's jump short, and he saw instantly that he was destined to join his now half-masticated horse in those shapeless jaws.

But in the next second, something amazing happened. There was a thrumming gust of air from just above Harry's head, and he felt blunt claws suddenly clasp his shoulders. With the force of a hanging and the noise of a cannon, he was snapped away from death and into the sky. He glanced down on the astonished face of his brother, watching as it shrank away with strange rapidity. Another purple-laced ball of perfect nothingness whizzed by two dozen feet away, and then Harry felt a wave of relief. He had been rescued.

Or had he simply been expelled into a new fire? Obviously, he had been taken into the sky by some sort of gigantic bird—though he had never heard of one like this. Its wings beat a steady, powerful rhythm, pushing air into his ears with each stroke. Perhaps he was to be food for its fantastic clutch. The bird climbed just a bit further, then the wingbeats became intermittent.

"Zat voss kloss! Ju ar very lucky I come!" It was a human voice, coming in fits and snatches through frantic panting and violent wind, from somewhere above Harry's head. Harry finally looked up, and his gaze found no bird above him, but a tangle of wood and metal, all infused with a strange glow. The voice had come from somewhere within what Harry suddenly realized was some sort of machine. Looking at his chest, he saw now that he was not gripped by any animal claw, but by strangely flat "feet," each an opposed pair of longish planks, reinforced with long strips of steel. They must have been articulated through some sort of joint, though as these were now directly athwart Harry's ears, he could only imagine their form. He could see where they attached to the main body of the machine, joined by a pair of sizeable leaf springs that were adding an extra layer of turbulence to his ride.

"Who are you?" Harry yelled through the ripping air. "Where are you taking me?" His mind began spinning new tales against his sense of safety—a human was no more necessarily an ally than a monster.

"My neem ess Lilienthal. Und I heff no idea ver I am takink you. But please to look for a high place, your rescue has left me very tired!" Suddenly they dropped violently in the air, and Harry feared he would be shaken loose from that mechanical grip.

"Lilienthal?" By god, an ally! A Union Engineer! Harry had never met the man, but had heard his name mentioned in some conversation about railway munitions. "This is simply miraculous, Lilienthal! I carry news from the front—we must make for General Grant's camp!"

"In which... direction?" came the reply, now ragged with exhaustion.

"You do not come from there?" Harry replied, crestfallen. He needed to find the General quickly—and not just for the sake of his allies and the threatened weapon. The feeling was beginning to fade from his shoulders. He also now remembered his foot, and looked down. The line of his boot cut off cleanly above the ankle, and there was no blood—but he knew that he had greater worries ahead if he did not get help soon.

"Zere!" the flying-man's shout was rich with relief, though when he wrenched the machine sharply left, Harry yelped in pain rather than thanks. But there it was below, unmistakable—a cluster of white tents, blue dots of men, and the flag of the Union visible in front of one particularly large tent. Even in his duress, Harry couldn't help but marvel—with no fires and surrounded by woods, the camp would not have been easy to spot from nearby ground, but from the air it leapt to the eye like a white banner. Then, with a jolt and lurch that nearly turned out Harry's stomach, they were swooping through the air and toward the ground with astonishing, if not alarming speed. They would arrive at that spot in mere moments, rather than the day it would have taken by foot from a similar mountaintop spy-post! Harry knew this thing—this mechanical bird, this flying machine—would be an immense asset for the Union. Lilienthal would be a household name throughout the world, forever remembered as the one who had given man the gift of flight.

He saw that Lilienthal, through whatever amazing system of direction at his command, was guiding the craft in the direction of an open sward amidst the smaller tents. There was no sign that they were seen from below—and after all, what reason had the camp's watchmen to look up, except to shirk their duty and contemplate the clouds!

Now they were mere moments from the ground, and the bird-ship's captain spoke again. "I must apologise, but zere is no other vey to do zis."

Before Harry could interpret this cryptic statement, he heard a loud click above him, felt the clamps around his shoulders loosen, and before a single thought could cross his mind, he was flying head over heels through the air. The sickening sensation of falling lasted only a few moments, though, before a violent, tumultuous impact.

Betrayed! Murdered! These half-articulate thoughts alone crossed Harry's mind in the final moment before his absurd death. All was black, and then...

More thoughts crossed his mind. Self-recriminations for revealing the camp to an enemy agent. Foreboding about the wholesale disaster that surely must follow. Then the question—why was he still thinking at all?

He curled his fingers, his toes. He felt them moving. He gently flexed his arms and legs, and was rewarded with more motion—and waves of dull pain that crested in his skull. He was, that pain told him surely, still alive. Groaning, he pushed one hand out in front of his face, and was momentarily blinded by a flood of daylight.

He realized what had happened—Lilienthal had dropped him into one of the larger tents, which had broken his fall as it collapsed around him. A great inventor and, it seemed, an excellent shot.

Of course, there was shouting and chaos now where the camp had been quiet and still only a moment before. Harry was still disoriented, wobbling more than a bit as he pushed himself onto his elbows. He found himself looking up into a dozen gun barrels.

"Identify yourself!" A nervous officer about Harry's own age gestured emphatically with his pistol.

"Lieutenant Harry McGee of the Union Army, under the service of Captain William Sherburne, carrying a message for General Grant!" Harry then began to gingerly and clumsily make his way to his knees, trying to keep at least one hand in the air—as if to ward off all the ammunition poised to dismember him—while treating the stump that had once been his foot with extreme care. The single bright side of the moment was the severe pain now inhabiting his entire body somewhat lessened that most dire one.

The officer was stymied by indecision, panting, his eyes panicked, but Harry's officious response seemed to work at least a slight calming effect. His breathe slowed as he babbled in confused fear. "McGee, where in the hell... what are you..." Harry could only guess at what these men had just heard or seen, and he felt great sympathy for the officer's confusion. "Men," the officer finally barked, "Secure this prisoner!"

Two soldiers stepped awkwardly over the collapsed tent—it must by pure luck have been empty. McGee, still kneeling, placed his hands behind his back to be bound, knowing that now was not the time for debate. "Sir," one of the men shouted dully, "His foot's off."

"Well, pick him up and we'll get him a medic," the lieutenant replied, suddenly dealing with something familiar. Not tying him up after all, the two men took his arms around their shoulders, and like a drunken spider the three extricated themselves from the tangle of canvas. The entire group then moved in mass through the camp, past dozens of tents.

Harry gasped when they finally stopped—in front of a pile of wreckage that must, only moments before, have been the flying machine. Wood and canvas and brass fittings were crumpled into a mass the size of a downed buffalo, with smaller shards and ends sprayed all round—including one intricate canvas wing, twice as long again as the machine's main body, angled into the sky against a tent that sagged beneath its weight. The other wing was nowhere to be seen. Harry's heart sank, knowing that surely the engineer Lilienthal had been dismembered with his machine—the method was worm food as sure as its product was kindling.

Then, the flap of a nearby tent swung open, and out strode the most dour and fearsome man Harry believed he had ever set eyes on: Ulysses Grant. The general had little of the shining glory of his namesake hero, instead moving his compact girth with grim deliberation. In sharp contrast to his men, who still sidled like frightened horses, the general gave no sense of his thoughts as he surveyed first the wreckage, then Harry. Finally, the grand old man addressed his new arrival.

"This news of yours had better be damned important."

Harry was struck half dumb. "Sir, how did you know–"

Grant leaped to within inches of Harry's face and cut him off with a roar. "By God, I'll ask the questions here! Even if I wasn't your superior officer, this…" Grant turned to gesture towards the wreckage, and his voice softened, suddenly nonplussed. "What the hell is this?"

McGee's head still rang, swam, wavered—but he tried to answer. "Sir, yes sir. There was a man, sir, an inventor, I was sent by Captain Sherburne at the magazine, they were sure to fall, and the Captain said you must know. He said there was a weapon, or some kind of… ammunition."

Grant's face was a collection of suspicious creases as he listened to Harry's stumbling report. He turned over his shoulder and shouted. "Bring out that Prussian pup!" He then turned back to Harry.

"Would this, then, be your savior?" Harry watched, thunderstruck, as two men emerged from a tent with a third between them—a teenage boy in suspenders and slacks, with a bare whisp of mustache and twisted, cracked spectacles strapped to his head by an awkward leather thong.

Harry immediately knew he had been tricked, and again damned himself a fool. "General Grant, I have been deceived! He claimed to be Max Lilienthal, the engineer! He must be some sort of spy, an imposter!"

Grant's face now took on a strange smile, but before he could speak the boy replied directly to Harry, his voice rigid with wounded pride. "I em no spy! Unt I did not say I am Max Lilienthal—zet is my father. I am Otto Lilienthal—and a loyal Union man!" He gave a slight bow.

Harry was about to reply again, but the slightest wave of Grant's hand cut him off. "This boy claims he invented the machine that… deposited you here. It seems unlikely, but intriguing."

Into the slight pause, Harry risked a comment. "It does not look like much now, sir, but it was an amazing thing. And the… boy… directed it expertly."

Grant sighed, and for a moment, chewed his mustache in a way that reminded Harry of Captain Sherburne. Crowns, Harry was learning, sat heavy.

"Release these men," Grant finally commanded. Then he took a second look at Harry and his missing foot. "Well, have the medic take a look at this one first. And find him something to walk around on."

Lilienthal, rubbing his wrists where they had been momentarily tied, came to stand alongside McGee, and faced Grant.

The General now spoke to both of them. "If the magazine really has fallen, then we need every bit of help we can get. We've got a matter of days to catch up with those Confederate bastards and take back what's ours before they turn it on us. The two of you know what's happened, and it seems you have some skills that may be of use. I'm hoping you're ready to get back into the thick of things."

Harry replied, hardly considering the details of what might lay ahead—or the fact that he was also speaking for an underage civilian who barely seemed made for the heat of battle.

"Sir, we wouldn't have it any other way."

To be continued…

I first met Steampunk Emma Goldman at the Steampunk World's Fair this year in New Jersey, where I had the privilege of attending a steampunk labor rally she'd organized. I've been a fan of the original Emma Goldman and Voltairine DeCleyre ever since I'd heard of them, so meeting her cosplay counterpart really made my day. Since this interview was conducted, the Occupy Wall Street protests came to life and Miriam has been arrested at least twice fighting against the system that empowers 1% of the population to rule the other 99%.

COSPLAYING THE GOOD FIGHT

EMMA GOLDMAN AND VOLTAIRINE DECLEYRE: STEAMPUNK'S OWN ANACHRO-ANARCHO-FEMINISTS

Illustrated by Juan Navarro

STEAMPUNK MAGAZINE: *So... Steampunk Emma Goldman and Steampunk Voltairine De Cleyre. Who are you?*

STEAMPUNK VOLTAIRINE DE CLEYRE: Does that mean who they were or who we are? Or who we are as them?

STEAMPUNK EMMA GOLDMAN: I was going to start with anachro-anarcho-feminism and go from there.

VOLTAIRINE: Sounds about right.

EMMA: We are anachro-anarcho-feminists, traveling through time and starting revolutions! In this time and general place, we tend to hang out at steampunk events, bringing a little touch of political awareness and involvement. We encourage people to use their knowledge of history to think about current, and future, political development. We've got a bunch of people with knowledge of the 19th century, who love to imagine the future. Who's better placed to think about politics than that?

My persona specifically is based on that of the historical figure of Emma Goldman, a feminist, anarchist, and activist active during the late 19th and early 20th century. She kicked epic amounts of ass. In real life, I am Miriam Rosenberg Roček, tall ship sailor, nanny, cook, and writer. I'm based in New York City; Steampunk Emma Goldman is based wherever she wants.

VOLTAIRINE: Voltairine De Cleyre was a contemporary of Emma Goldman's. Like Emma she was a feminist, anarchist, and activist. She got into a lot less trouble with the law, and they could never agree on individualism versus collectivist anarchy, but she wrote a lot of essays and did a metric tonne of speaking engagements. As a persona, as a time-travelling steampunk, she's got access to anti-depressants we decided has made her relationship with Emma Goldman much better than it was in real life, and they gad about Time, as Miriam said, using the past to look at the present, and as a tool to imagine, and work for, a better future.

In real life I'm Elizabeth Burns; writer, shopgirl, assistant. I'm based in Toronto these days where I'm secretary to the Toronto Steampunk Society (I know, we're actually very organized, not anarchical at all), but Miriam and I went to university together, which is how we know one another.

SPM: *What made you decide to adopt historical personas, and these in particular?*

EMMA: I'd say the reason I picked Emma Goldman, and the reason I decided to do a persona at all, was that I had noticed that while steampunk spends a lot of time, as a culture, admiring and remembering great artists, writers, and scientists, we don't talk much about the activists of the nineteenth century, and the amazing work they did and insane lives they led. This is a way for me to remind people that activists, and Emma Goldman specifically, are really worth remembering; they're not only a huge part of the story of the time we're all interested in, but what they did influences and impacts us today.

VOLTAIRINE: Miriam was raised by anthropologists, and went to a Quaker school, which I think explains everything. I mean, it doesn't explain her hatred of penguins, but it covers the activism awareness.

EMMA: Thanks, Elizabeth. That really clears things up. Anyway, Emma Goldman appealed to me because she was such a dramatic figure, and because, while I don't agree with everything she stood for or everything she said and did, I feel like the world was a better place for her being in it, and the world needed, and continues to need, people like her. Roleplaying her allows me to put a little bit of her revolutionary fire into steampunk, and into the world, without actually becoming a revolutionary. It seemed like a good tool to help me politicize steampunk. I actually only intended it to be a one-off thing; I dressed up this way for the Steampunk World's Fair in 2011, and the reaction to it was so overwhelmingly positive that I went home, made a facebook page, and started planning Steampunk Emma's next event.

VOLTAIRINE: Don't let her fool you, she picked Emma because she likes her style. Not that she's leaping up on stages to thwack people with horsewhips (she might though), but Miriam and Emma have a lot of the same, "It's interesting that you have that political opinion, but I think you'll find you're wrong. Yeah, sure, interesting point, but also? Wrong," way of dealing with bigots and those unwilling to see their privilege. Revolution, dancing, and a surety that anyone who argues against her is probably an idiot. It would be very hard to put up with if she wasn't usually right.

EMMA: There is that. I think the really critical thing, though, was that Emma Goldman was just so amazingly f-ing cool, that being able to dress up as her is a little like being a kid, tying a towel around my neck and pretending to be Batman. Plus, I found a pince-nez at a flea market, so it just seemed like the thing to do.

VOLTAIRINE: For me, I was never politically aware until I went to university and got a nasty shock. I'd been naïve, I'd been to an all-girls school in England that led me to believe the world was wide open, and while I was aware of class issues, going to the States was eye-opening. Miriam was pretty instrumental in moving me along from being angry about the unfairness in the world to actually doing something about it. I went to my first protest with her, and cringed about making a scene the whole time, but that was the start.

EMMA: Ha, I remember that. Elizabeth was like, "What are you doing? I'm way too Canadian to be rude to a police officer!" And I was like, "I'm not being rude, I'm telling him he's full of crap and demanding to know what law we're breaking. Politely." We didn't get arrested though, which was good.

VOLTAIRINE: Anyways, when Miriam asked if I wanted to join her cosplay, I fired up a search engine and started looking for someone who fit. Voltairine really spoke to me. She wasn't like Emma, she didn't get arrested, didn't capture the public eye, but she was a writer, an agnostic who wanted to be an atheist, a staunch feminist, and a depressive who spoke for the rights of the mentally ill, though it wasn't something really understood at the time. Reading her essays made me feel like I understood her, as much as anyone can, I suppose. She didn't get on personally with Emma, but they respected one another, and it seemed to me that if Voltairine had been given proper mental health care, and was able to be a happier, healthier person, she and Emma would probably have been quite close. Also, I can get my hair to do that weird curly thing she had going on.

It's been a rocky start; Canadians tend towards political apathy. Politicians are largely ignored/ignorable and we have gay marriage, health care, access to abortion, unemployment benefits etc., so we're a little arrogant about how well we're doing, especially in comparison to the USA. It's been hard to get support. Plus, I'm still a little nervous about speaking up—not to people I know, but in public. It's nice to have Voltairine to hide behind (also Emma), just a little, and remember that she was a change for good, even though she didn't love the political spotlight either.

***SPM:** Ms. Goldman, you were one of the main organizing forces behind the labor rally at the Steampunk World's Fair this year, am I right? Can you tell me about that?*

EMMA: Yes, that was actually the first time I did this character. I had been talking, or perhaps, mutually ranting would be the better term, with my friend Leanna Rennee Hieber about Scott Walker, the decline of unions in America, and the ridiculous anti-union rhetoric spouted by conservative politicians. Somewhere along the line, we decided that the big problem was that people had forgotten their history, and that it would be a great idea to hold a 19th-century-style labor rally to remind them of it. This sort of things seems like a logical leap when it's two in the morning and everyone's wearing corsets, I promise.

I ended up organizing the event, as Leanna is a very busy person. I encouraged people to come out in support of organized labor past, present, and future, real and fictional, and people responded with delightful enthusiasm. I'd say my two favorite bits of that rally were the young kids who dressed up as newsies from the Newsboys' Strike of 1899, complete with signs condemning Pulitzer and Hearst, handing out pamphlets about the rally in place of newspapers, and the tongue-in-cheek counterprotestors, who showed up dressed like a 19th-century gentleman, waving signs that said "Get back to work, peasants" and "why work when you can inherit?" It was hilarious and fun, but more importantly, it got people talking and thinking, not just about organized labor, but about the idea of using steampunk as a political tool. I'm going to keep putting on events that are fun, entertaining, and political in nature; I don't see why fun should detract from political usefulness, or why political relevance should detract from enjoyment.

SPM: *What draws you all to steampunk?*

EMMA: Do I lose points for shallowness if I say that what first grabbed me were the clothes? I really, really like the clothes; Elizabeth awakened my goth side when we were both in college, and then steampunk helped me expand on that. On a less shallow level, I love the blending of history, fantasy, and science fiction. I love that we can play with history, use it to flavor everything we do. I love the DIY culture, and the repurposing of old clothes and technology. I love Nikola Tesla, Jules Verne, Abney Park, and absinthe.

VOLTAIRINE: As a side note, I've never tried absinthe because I hate the taste of aniseed. I feel like I'd have to drink it in an old-fashioned bathtub while having a vision and thinking about Jack the Ripper. I also feel like I've seen *From Hell* too many times.

But more to the point! I was a big goth in high-school and I was always drawn to the theatricality of dressing up in different styles within the genre, one of them being a more Victorian sort of look. I also studied Victorian literature and history quite extensively at the time, and then continued when I went to university, expanding into the American West, and New York. It was also then that I was free to do whatever I wanted since everyone thought I was weird anyway—wear something other than jeans and you get asked if you're going somewhere nice, on that campus—so why not go all out? Having someone else who was just as interested in exploring history, sci-fi, and the limits of how long you can stay awake while abusing caffeine and light-boxes, made it all the more fun.

We had foam sword duels at 4am in a White Hen, celebrated Jack the Ripper's double-header night (for want of a holiday), dressed like bandits, read each other great works of literature while we made feeble overtures at cleaning our apartment, wore bustle skirts and dread-extensions, researched everything that caught our interest, shared factoids we'd both cribbed from Horrible Histories, and generally made asses of ourselves in public, and had a wonderful time doing so...

Miriam and I lived together for two years, and I think we just sort of fumbled our way into the genre through our shared interests. We read a lot of the same sci-fi and fantasy books when we were kids, so there was that to start with, and as I said, we both shared a love of history, literature, and exploring alternative or hidden stories within the past. I spend a lot of time online futzing around, so I'm sure I stumbled across some link or other that clued me in to the emerging genre. Turned out what we were doing bore a strong resemblance to this steampunk thing. The rest was history. Or possibly the future. This time travel thing makes tenses very difficult.

It's also a wonderful social network. I moved to Canada about three years ago and I didn't know anyone. So I looked up the local steampunk society and one Tinker, Trade, and Tea later, and I had people who were happy to show me the city, people to lunch with, see movies with, hang out with. It really made my transition much easier than I had ever hoped, and I think at its heart that's what makes this genre so remarkable is the people it brings together. There are so many talented, creative, fun steampunks and we all share overlapping interests, so there's always something new to learn and explore. And I mean, where else can you tell a hard-science physics joke, in a pub, to people dressed like sky pirates, and get a real laugh?

The Paraclete of Pierre-Simon Laplace

by Jamie Murray

LUC TRUFFAUT RECENTLY PUBLISHED HIS BOOK *"THE BROADER IMPLICAtions of Mécanique Céleste"*. The details of this book are well known and it is scarcely necessary for me to discuss them here. However during my reading I was snared by a naively clichéd footnote in chapter 26 about Laplace's speculation that "an intelligence which could comprehend all the forces by which nature is animated and the respective situation of the beings who compose it and intelligence sufficiently vast to submit these data to analysis it would embrace in the same formula the movements of the greatest bodies of the universe and those of the lightest atom; for it, nothing would be uncertain and the future, as the past would be present to its eyes." In summation, if an intelligence knew the location of every molecule in the universe it would be able to *pre*dict and *retro*dict any event across the life of the universe. In a book with pretensions of definition on the work of Pierre-Simon Laplace, I could scarcely believe that the most profound affect of Laplace's influence would be reduced to the following footnote on page 921:

> "The Ukrainian mathematician and physicist Konstantin Tcherevitchenko spent much of his academic life studying the theories of Pierre-Simon Laplace and was notable for developing a peculiar cursive script that could be used in lieu of mathematical notation. This script was not widely, or even scantily, adopted by others in the field but it is still believed by some that it brought greater clarity to mathematical formulae."

The tragedy of Tcherevitchenko's work being reduced to such a brief condensate could scarcely be believed had it not appeared in print for all to gaze upon like children around a doormouse cadaver. Tcherevitchenko was undoubtedly favoured highest amongst all Laplace's academic progeny and was the true heritor of his mathematical dynasty.

Konstantin Tcherevitchenko was born in Eastern Ukraine to a wealthy family. His father, Nicolai Tcherevitchenko, was the Royal

Malacologist to Tsar Pyotr IV and Tsar Alexander I. He would later develop a paradigm of of scientific management called "On the correct appropriation of burdens to their beasts" which was grounded in physiognomy rather than engineering but was still published nearly a century before the better known work of Frederic Taylor. Tcherevitchenko's mother, Irina Tcherevitchenko, franchised herself with her vanity and often made the outlandish claim that she owned a garment or bauble from the hide, calculi, or ivory of every animal on the Ark. This changed when she began a new beauty treatment involving the complete immersion of her head in coco flowers and ermine milk. Unknown to Irina Tcherevitchenko, the beauty mark on her cheek that she coveted so dearly was a commensal parasite which promptly drowned in the pomade; leaving Irina to mourn the death of her vanity's testimony for the rest of her days. As a boy Konstantin Tcherevitchenko was considered very anomalous by his nurses and teachers as he suffered from unprejudiced pica and yet would refuse to have his tea sweetened with anything but damson jam. He was also believed a mute until the age of thirteen but later revealed to his parents that he merely had a very eloquent and finely bred inner monologue. During this period of inward soliloquay he read *A Philosophical Essay on Probabilities* repeatedly and ecstatically, honing his love mathematics and unearthing new curiosities all the time.

The relevance of Konstantin Tcherevitchenko's life begins when he was still a student at the Saint Petersburg Imperial University. Kolyakov (1879) explains that Tcherevitchenko's studies began to include the work of Ibn Khaldun, Alberuni, Ibn Turk, Alhazen as well as Laplace and he became fascinated by the way they would describe the operations of the universe in a flowing, joined script. Tcherevitchenko felt this described the nature of force, curves and logic far better than the block like, geometric writing of Roman or Cyrillic script. Within a few short months he was writing all of his work in ancient Arabic and only converted them into French or Russian for the convenience of his professors. When Tcherevitchenko was nineteen years old it came time for him to write his doctoral thesis. By this time the language he expressed his thoughts in was no longer the Arabic of ancient times but had taken on the role of pure mathematical notation. Tcherevitchenko would write his doctoral thesis on his new mode of notation and numerals and, as an example of its use, he would translate an entire volume of *Mécanique Céleste*.

And yet this language was no rude inventory of integers. Tcherevitchenko's language could describe sectors or coordinates, equalities and wholes, fractions and speeds with greater precision and created distinctions that had hitherto gone unnoticed by generations past. The most notable example was that 0 and 1, as with all numbers, were no longer mutual exclusives of one another but were vowel-like concepts that could be described across an ordinal gradient, and expressed in many ways, each with its own subtle nuance.

Tcherevitchenko completed his thesis and submitted it to his two doctoral advisors. One of Tcherevitchenko's doctoral advisors, Evgeny Vasilev, was chief glutton of the gourmet society during his own undergraduate pupation and was considered rather superlative with the blunderbuss and the lanner falcon. He was made chief glutton after serving up a French service banquet he called *Melange Gris*; made by running an alternating electrical current through a river and serving up all the "strange flesh" he skimmed off the surface. His spatchcocked jenny hanivers on a bed of "something like escargot" and his gull stuffed with toad rollmops were also highly esteemed; although his choice of digestif liqueurs was considered very poor in both instances. The other, Arkady Kirsanov, was responsible for introducing a variety of French diseases to Georgia which he mistakenly acquired during a trip to London when he was given inaccurate directions to his hotel.

Vasilev and Kirsanov were not impressed by his thesis. They considered his work to be superfluous and brought very little to the current body of knowledge in mathematics. His thesis would have been failed but they recognised that a great deal of toil had been bestowed on it and that it was otherwise a well written, albeit decorative, piece of work. So he was awarded his doctorate in mathematics. The unpopularity of his work meant that it was never published and the practice was not adopted by any other physicists or mathematicians, so it sat on a hypothetical shelf for the next thirty years or so. Tcherevitchenko was unwilling to become a scholar due to his disillusionment, so he spent most of his life working as chief tactician for the Romanov House Cavalry and the 12th Tsaritsyn Grenadiers. Though Tcherevitchenko found the work to be banal, it would later prove serendipitous.

Kramskoi's *Russian Historical Affairs of Civil Disobedience* mentions Tcherevitchenko and that he was sent with the 12th Tsaritsyn Grenadiers to pacify the Siyowpidi tribe that had been invading the Russian provinces and claiming them as their own. The Siyowpidi fought with a blatant disregard for tunication and their orphan tactic was to use slings and atlatls in a highland charge, after the fashion of the Jacobites, before pressing closer with percussive weapons edged with chert.

Conversely the Tsaritsyn Grenadiers carried plundered Ottoman repeater rifles with piston driven bayonets. Their satchels were filled like apple bushels, with bombs no longer garbed in vulgate shrapnel; but live cataplectic hornets, arranged in neat rows like horrifying embryos in parade formation. The 12th were truly guildsmen within the noblest profession and were veteraned with distinction at the battles of Pyeizley and Oshvaya, the siege of Elephantine Castle and the The Last Stand at Blayzhvid Ploshchad. They were fortified with Crimean sky corsairs and velocipede-mounted Hun mercenaries. The Siyowpidi could have outnumbered the Tsaritsyn Grenadiers many times over and still given them little cause for consternation.

And yet this was exactly how they felt.

The reason for this anxiety was that the Siyowpidi were a shepherd people who lived in the mountains around the Vovtuzenko Highlands where a morbid affection known as marmoreal lung had killed off so many in generations past. The Austrian physician, P.S.D. Crönenberg described it thus:

> "Marmoreal lung, so named because of the charonic white colour and marbled aspect of blood vessels in post mortum specimens, was of a similarly horrific species to miner's lung although their causes are peculiar to one another. Marmoreal lung was originally seen only in goats but spread to the human population via an, as yet, clandestine route. The bacteria cluster around the alveolar endothelial cells in the organ, respiring with the inspired oxygen and increasing in numbers. The victim will breath more rapidly in order to compensate for the reduced volume of oxygen reaching the blood. The increase in oxygen concentration in the cavity of the lung will increase the metabolic rate and therefore the rate of division of the bacteria. The tachypnoeic rate of the infected individual will now be the basic requirement for the body's oxygen tithes although the demands of anxiety and hypercapnia, such as increased activity, ventilation and hysteria, will further increase the body's oxygen demands and thereby the repetition of bacterial meiosis. Sufferers will complain of not being able to sufficiently insufflate the chest and will often complain of sensations not dissimilar to drowning. When someone is infected, they will usually be rendered prostrate within minutes but it will take some time for death to occur. A matter of days rather than hours would normally be expected."

The Siyowpidi, through a fashion of hominid husbandry, had become resistant to the illness and so proceeded to claim the land emptied by marmoreal lung as their own. The Siyowpidi eventually reached obshchina that were still in a state of habitation, at this point reticence lost their favour and they refashioned their migration into an invasion. The Siyowpidi were poorly equipped and lacked the means of modern warmaking such as powder and shot weapons and could not afford to purchase them from others. They realised that they should rely on the only means of survival they had hitherto found successful. They willfully infected prisoners of war with marmoreal lung in order to use their bodies as nurseries to ripen the infection. When the prisoner approached death the Siyowpidi would collect the expectorate from their lungs. This expectorate would then be smeared upon their bodies and the blades of their weapons. The Siyowpidi would charge across the battlefield, beatified in their contagion, and killing with their presence as much as their battle fury. The Tsaritsyn grenadiers' valour, which remained redoubtable when fear was presented upon the point of a pike or the curve of a bullet, was overpowered by their squeamishness. Such was their horror of catching the dreaded disease or being held as uterine captives.

Despite their rude weapons and indecent armour, the Siyowpidi had proven themselves against every army that had tried to stop their expansion. The sergeants and ensigns refused to engage the Siyowpidi and instead insisted that they fire Parthian volleys as the Siyowpidi approached in order to lead them into an area mined with petards. Tcherevitchenko calculated that more of the Tsaritsyn would be killed by missile fire and the Siyowpidi would still engage them in a direct assault. Tcherevitchenko had developed a bespoke maneuver for this battle known as "The Helicon"; and whilst it involved a direct frontal assault, it would completely overpower the Siyowpidi with minimal casualties to the Tsaritsyn. Though the generals of the army held the opinions of their soldiery in contempt more than esteem, without them they would have none at all; so they ordered Tcherevitchenko back to his tent to devise another solution.

Tcherevitchenko sulked on the thought that it would be so much easier to deploy soldiers that had been trained since birth. Their minds would be limed and tanned against any such distraction. At this point he suddenly remembered Pierre-Simon Laplace's famous thoughts on the uncertainty of the past and future. Tcherevitchenko's head blended the two thoughts into potent theopneusty. Just as a soldier's present state is the effect of its anterior or infant state and is the cause of the one which is to follow, so it would be with anything else.

Tcherevitchenko would need solitude and tranquility for his work so he saddled his horse and stole another with the intention of saturating its burden with an entire regimental ration of wormwood liquor; his mind would, after all, need to be suitably primed if he was to calculate every event in creation. He travelled for six hours and then decided to stop at the nearby Cassandra Inn.

He ordered a room, opened one of the cases of wormwood liquor, ordered the first of several plates of pelmeni and spiced cabbage, and began working immediately.

Many theoretical models of the infant state of the cosmos already existed such as Jevon's intussuscepted universe, Gray's schwa universe, Carrol's polylobular nucleus universe, Adams's xioid universe and Lloyd's Blivet universe. Tcherevitchenko found conventional numbers grossly inefficient so he decided to write down his formulae in the very notation that his professors had thought so little of. Tcherevitchenko's revelation was that all one would have to do in order to determine the location of all the molecules of the universe in the present would be to determine their location at the universe's infant state and then complete the algorithm of its existence to the present. Through trial and error Tcherevitchenko managed to select the correct shape of the universe, which was Lloyd's Blivet model cosmos, before he determined the location of the molecules at the beginning of the universe; from the glaring lights of the integers and the shadows in between them, Tcherevitchenko was able to make out the first born molecules of the universe, large and fat like dumplings in broth. Tcherevitchenko would surely have been ecstatic at seeing both the universe and his great work laid out before him in the peculiar flowing script of his own devising.

The implications of it could not be slighted. No longer would the past be the subject of historians' capricious interpretation or scrutinised in the merely cross-sectional form of the social anthropologist. It would cease to be a fragmented extrapolation from lumps and chips an archaeologists' catalogues and philologists would no longer be needed to reverse engineer ancient conversation from poems and letters. Tcherevitchenko's pogrom was absolute; there would be no more speculation on future events and even etymology, heresiology, genealogy and theology would forever be cast into obsolescence.

The data was initially nebulous and Tcherevitchenko was unsure of what he was describing, this was because he had very little knowledge of geology and what he was observing was the formation of the Earth. Gradually he began to make sense of what he was observing like the gloss emerging from shoes as they are polished. When life combusted upon the Earth it was like a flame dancing in praise to all the song that would ever be sung; he saw bacteria syndicated into a Gastrea, princely lizards ruling the Earth until loquacious little apes acceded their thrones and existence while monolithic trees watched in silence. Tcherevitchenko watched with lamenting eyes whilst mankind divided its labours and then weeped in its multiplicitous sorrow as it divided itself by the same action. With louped scrutiny Tcherevitcheko could make out individuals, such as Imhotep standing at the the right hand of Djoser, Archimedes unleashing his magic of levers and optics upon the Roman navy, lycan madmen reaping out nomadic kingdoms with the edge of an axe and Robert Burns cherishing his joyous memories of Agnes M'Lehose that circumstance had painted with forlorn colours.

He began to decipher what he was seeing in the future as well. The future remained equally enigmatic until the universe had shrunk to its present size and the branch of the tree of life had receded to the perch on which mankind currently sat. Tcherevitchenko watched humankind leaving the scorched planet that had been their cradle for so long only to meet their star-born peers. He observed the first machines built in the image of the human mind and he saw two kingless empires threaten each other with weapons that none could ever use. He watched as millions of men were sent to the barbwired and duckboarded hell where youth and laughter go; all the while poets wove ugly tapestries with beautiful words. The past and future slowly converged upon the present, the past slowly approaching its omega and the future its alpha. This was when Tcherevitchenko found his most chilling discovery.

His death.

What's more was that he was to be found dead the very next morning. The thoughts that went through Tcherevitchenko's head are unknown, if he bore the grave news stoically, if he challenged fate in some unsuccessful way or if he accepted fate's charge rather than waiting for the sword of Damacles to fall. Whatever his feelings were, he was found dead the next morning; a Zapaporozhtzi sabre that was not his own had punctured his vitality. What Tcherevitchenko deigned to tell the world was the occasion for his death and of the indecipherable writing in his room. Writing that was penned in ink and chalk as well as being scratched or embroidered directly on to surfaces and covered the walls, sheets of paper, upholstery, floorboards and even etched into his skin with a sharpened fountain pen as though he were some antipodean junker. Everything else remains unknown.

Misrepresented by history, misdiagnosed as a boy, misjudged as a student and mistrusted as a tactician. Tcherevitchenko suffered all of these things and still became the sorcerer-mathematician prophesied by Pierre-Simon Laplace. The writing on the walls of that single room of the Cassandra Inn was transcribed and most universities across the world hold copies of it, but the spurned doctoral thesis which can translate it has never been found. Could this be fate's wry and cruel sarcasm that taunts her playthings like an imp holding grapes over the head of Tantalus? Or is fate actually revealing her grace to us all? Would that knowledge, avariciously hoarded by oracles, be a greater plague than any cast upon a people? Our lives would be fertile with knowledge but arid of pleasure; we would have no surprises in our future and no nostalgia in our past. So it must be put to rest as merely an ammonium phial for our curiosity; for only one man, whose wax bound wings melted in the warm rays of time, can ever know.

THE MAKER

AN INTERVIEW WITH THOMAS WILLEFORD:

Conducted and illustrated by E.M. Johnson

If you've been looking for the embodiment of what it means to be a professional steampunk, look no further than Thomas Willeford. One of the most recognizable individuals in the steampunk community, with his red coat and mechanical arm that looks as if it's straight out of a film, Thomas has made a name for himself creating amazing pieces of wearable art that blur the line between costumery and pure engineering. With over 20 years of experience as a maker, Thomas is the mastermind behind the company Brute Force Studios, and author of the newly released book Steampunk Gear, Gadgets, and Gizmos: A Maker's Guide to Creating Modern Artifacts, *where he shares some of his secrets with those of us who would like to make some steampunk gear of our very own. Thomas was nice enough to take time out of his schedule as maker extraordinaire to talk with me about his work, what it takes to be a maker, and why corporations have failed to capitalize on the steampunk mystique.*

E.M. JOHNSON: *First and foremost, I have to ask: why steampunk? With degrees in physics, Victorian history, and art, it seems that you and steampunk are a match made in heaven, but what was it about steampunk that captured your attention and inspired you to become a professional steampunk?*

THOMAS WILLEFORD: Growing up with my grandparents in a big Victorian house (they truly were mad scientists) kind of gives it that sense of inevitability. I always loved Victorian science fiction as a kid. In my college days, it was my rebellious nature which made me go for Victorian and steampunk. All my friends were pretending to be rebellious by following Medieval Studies in school. In 1988, I started playing *Space 1889* (yes, I was one of the few people that actually played the game) and I was pretty much hooked.

E.M.: *Where do you get your inspiration? It is mainly from books and media, or do you find most of your inspiration when you are scavenging around flea markets and antique sales?*

THOMAS: Sometimes it's books, media, and bits of junk lying around the workshop. I also have an entire, fully-developed steampunk universe in my head and I want to make things that fit in it.

E.M.: *As steampunk continues to gain popularity, many people are led to believe that steampunk is simply a particular style or fashion aesthetic, a mindset which you describe as "cog on a stick" in your book. You go on to point out that you can't simply slap a gear on anything and call it steampunk. What do you feel is at the heart of steampunk art and fashion, and why does true steampunk art and fashion go above and beyond the simple "cog on a stick" mentality?*

THOMAS: Of course you can slap a gear on something and call it steampunk—and I'm allowed to laugh at you for doing it. Again, the aesthetic works best when it at least appears functional. To me the heart of steampunk art and fashion is the adventure. There's the guy who actually goes on an adventure and flies an airship. The cog on a stick types are just buying the postcard.

E.M.: *Many companies have tried to jump on the steampunk train during the past few years, mass-producing everything from steampunk gear to jewelry to clothing. As a maker, how do you feel about this attempt to mass-produce steampunk, and why do you think that these attempts have mostly failed?*

THOMAS: I have so far found their attempts actually quite humorous. Why is it failing? Easy. They're not actually asking anyone in the steampunk community to design anything. They're telling their designers "Go out and find me something steampunk" and they're getting watered down herbal tea version at best ("You say 'herbal', I say 'No, thanks!'").

E.M.: *Do you think that there is a way to make steampunk more accessible to those who might not have the time or the money to invest in hand-crafted items (especially if they are new to steampunk) or do you think that the very foundation of steampunk art and fashion is unable to be divorced from the ingenuity and craftsmanship of those who create it?*

THOMAS: This gets into classism. I considered this question for a while because oh my god, does it sound like a trap... but I'm going to step right into it. As with anything worthwhile, you need either time or money. If you take the time, you can save loads of money and make beautiful things. Conversely, if you've got the money, you can save loads of time and buy those beautiful things. Steampunk is not for those of the Wal-Mart mentality.

E.M.: *After reading Kaja Foglio's (of* Girl Genius*) introduction to your book, it becomes clear that the steampunk community encourages and appreciates making and craftsmanship more than many other communities. While it might seem like asking if the chicken or the egg came first, do you think that creative people are drawn to steampunk, or that being a part of steampunk culture makes people more creative than they may have been otherwise?*

THOMAS: I personally have the very unpopular view that you cannot make someone more creative. You can only show them tools and venues to explore their creativity. I keep meeting loads of people who make incredible devices. You walk up to them and say "That's a great steampunk piece you have there!" and they say "A what?" but once they know this community exists, they're quite often hooked. Steampunk can be very inspiring toward people who are already creative. Again, TRAP!

E.M.: *Why do you feel that today's technology lacks beauty, when much of the technology in the past still had a certain kind of workmanship and beauty to it? Do you think that the modern penchant for cookie cutter technology is part of the reason that so many individuals are drawn to steampunk?*

THOMAS: Modern technology is bland because it's all about selling to the lowest common denominator. There are more people without taste than there are with it, so it's much easier to sell to the lower part of the pyramid. I think that people are drawn to steampunk because there is not a lot of fare for those higher up on that pyramid. They see this and they're kind of hungry for it.

E.M.: *If you could recommend any project to start with for those who are fairly new to making, what would it be, and what materials should they keep an eye out for?*

THOMAS: I would say the goggles in the first chapter of my book are a good first project because they give you practice at a lot of skills you will use in other projects: leather working, metal working, and even (ew) plastics and you'll look really cool wearing them while you make your other projects.

E.M.: *Is there any specific media or literature that you would recommend to those who are trying to get a good grasp on the steampunk aesthetic, specifically the artistic part of steampunk?*

THOMAS: The original *Wild Wild West* from the 1960s (it's even available on DVD), Greg Broadmore's *Dr. Grordbort's Contrapulatronic Dingus Directory* and *Victory!*, the "steampunk" 12-part comic book series from Cliffhanger, and definitely Phil and Kaja Foglio's *Girl Genius* are all excellent places to start (some of them are even good places to hang around for a while).

E.M.: *What would you like to say to those who are thinking about pursuing making as a career? Any tips, tricks, or warnings?*

THOMAS: Treat it like any other job. You go to work in the morning, you make your stuff, you have lunch at noon...it's a business first. Otherwise, it's just a hobby.

E.M.: *Is there anything else that you'd like to say that wasn't asked? (Victorian party tricks, interesting physics facts, favorite movies, etc. are all fair game!)*

THOMAS: One of my favorite things to look up and find is "real steampunk": the first contact lenses from the 1800s, Victorian-era prosthetic limbs that had functions, the airship in the 1895 "Spirit of Transportation" sculpture at the Philadelphia train station. These things can be truly inspirational. It's really great to see how close we came to steampunk being our actual past.

You can see more of Thomas's work and purchase an autographed copy of his book on his website, BRUTEFORCELEATHER.COM.

Manny Aguilera
MannyCartoon.com

HOW TO SNEAK AROUND

by Wes Modes
Illustration by Manny Aguilera

In the US, in the post-September 11th universe, people are much more suspicious, more questioning, and more likely to arrest first and ask questions later.

For you, that means you need to exercise a little common sense, a little more caution, and be perhaps quicker to admit that you are just some dumbshit looking for thrills. As always, don't run from authority, don't make things worse—play respectful, have your ID, and be ready and willing to bear the consequence of your mischief. Other than those heavy words, have a blast!

One, two, three. I count the steel rungs as I climb below street level. I lower the grate over my head back in place. Four, five, six. My boots squish down on the muddy, leafy floor. I look around. Tunnels disappear in the murky distance in two directions. I am standing 15 feet below the university. My heart is racing double-time. I am equipped with my trusty flashlight and waterproof boots. Two minutes ago, I stood looking down through the bars of an unlocked grating covering a hole that descended into the depths. I asked myself: Well, why the hell not?

This is all about the fine art of getting into and out of places you don't belong. Why? For some of us, just the thrill of peeking behind the scenes of life is enough. Add to that the challenge of evading the law and pushing personal and societal boundaries. Toss in the opportunity to indulge a childish urge to run around like Indiana Jones. And it's too much to resist.

Sneaking around can be more than cheap thrills. It's a useful skill for journalists, photographers, street artists, activists, survivalists, travelers, urban commandos, trespassers, and adventurers.

Here's how to do it.

Becoming Invisible

Michael leads me into the basement of the MAC, past all the practice rooms, through the property department, past the sunken orchestra pit, up some stairs. We pass several people and we steam by them without looking back. They don't give us a second glance. Across the empty stage and through the sets waiting in the wings, we enter a tiny old crimson-lined elevator. The elevator indicator keeps going up, up, up, six, seven, eight floors. The elevator dings and stops at 10. Michael pulls open the door and out we step into…

Nothing. Darkness. Slowly my eyes adjust. There is light coming from below. I look down. We are 100 feet above the stage! I can look down and make out the network of catwalks above the stage. Hundreds of ropes and cables descend from pulleys into the distant depths. I get that funny feeling that makes me think if I get too close to the edge, I may throw myself over. I am dizzy from the height. Michael is smiling.

"Wait until you see this," Michael says. "You won't believe it."

Look and act like you belong. You must fit in. If you are playing an inspector, you must act like one. Look around as you go, maybe nodding, maybe looking at a clipboard. If you are an office worker, you must walk looking straight ahead, drone-like to that business meeting elsewhere. You must walk like you belong there. You must think like you belong there.

Keep moving. As long as you are moving, people assume you know where you are going. The moment you stop, you invite people to ask the dreaded question, "Can I help you?" As you walk through a lobby, steam right past the security desk with a nod. If you reach a dead-end or locked door, act like you *meant* to reach a dead end-turn around a keep moving. Only when you aren't observed can you take the time to regroup or plan.

Take advantage of people's hesitation to question strangers. If you keep moving, you are out of their sight by the time they make a decision, saving them from deciding at all. Out of sight, out of mind.

Avoid eye-contact. Eye-to-eye contact, especially with security people invites questions. If you look at people, flash them a quick, confident smile as you pass.

Don't act suspicious. Stealth is good only if you know you won't get caught. If you are found slinking around, it will be a lot harder to pass off the story that you were just looking for the bathroom. Checking locked doors, hiding, running, or looking around nervously will make you look suspicious as hell. If you have to be a suspicious character, make sure you are not being observed doing it.

Wearing A Disguise

We were wandering around Walt Disney's Magic Kingdom where we weren't invited.

While two employees stood around doing their thing, we looked at the forbidden secrets of Tomorrowland, gesticulating, pointing, and nodding in an inspectorial sort of way. We were dressed as nerdy techs and we fit right in as we walked with confidence around the back lot of Disneyland.

Use a disguise. When was the last time you questioned the intentions of the UPS man, or the Federal Express woman? Did you ever wonder whether someone wearing a hard hat and an orange vest was *supposed* to be putting cones in the road? A disguise can be remarkably effective. When you put on a uniform, you don a mantle of respectability and responsibility. People *expect* someone with a hard-hat and a clipboard to casually brush past the receptionist to perform an "inspection."

Take a look at how simple most uniforms are. Put on a dark brown shirt and pants and grab a package with the right address on it and you become a UPS delivery person. Put on a hard-hat and some rugged clothing and you become a construction worker. Put on a suit and you become a business person beyond reproach or question.

Blend in. Another disguise is the one that simply blends in. In an office building, a suit and tie is the perfect uniform of conformity. On a job site, a hard-hat does the trick. At a formal function, a tuxedo is your ticket.

Dress up. Dress one step more formal than others, and you will move with impunity. Everyone will assume you are another step up the organizational ladder. No one is willing to risk embarrassment to ask aloud what you are doing there. Look important and everyone will assume you are.

Getting Caught

The sun was just setting in Sante Fe as we reached the yards marked with the blue Santa Fe cross and signs reading, "Santa Fe Property—No Trespassing—Violators will be prosecuted." The Santa Fe yard looked much more secure than the Southern Pacific yards we'd been in. Barbed wire topped chain link all the way around.

We walked in one of the main gates and down a road to the yard. We were joking about what we'd say if we were stopped. We looked so obvious with our giant packs, boots, rugged, bulky clothes. We were sure we'd get arrested or at least thrown out of the yard. "We're, uh, looking to get across the yard. To, uh, Washington Street," Philip said when we were questioned by the first car that drove by.

Getting caught doesn't need to spoil the fun.

Have a good story. Make up a lie that explains why you are where you shouldn't be. Or tell the truth if you think it'll work better—for instance, you're just curious and wanted to see what was in here. But make up your story before you go in so you won't be doing that implausible stammering and stuttering when you get nailed.

Maintain plausible deniability. A good story should be believable but unverifiable. For instance, saying you are a friend of the owner may put you in a very uncomfortable position when the security guard calls your bluff. On the other hand, if you claim you are a freelancer and were told over the phone that it was okay to enter the building, your story is difficult to disprove. This story may not give you carte blanche, but you may only be ushered off the premises rather than arrested. If plausible deniability worked for the Reagan crowd during Iran-Contra, it may work for you.

Play dumb. Don't be afraid to plead stupid. "I'm lost," is probably the most effective plea. Few people are going to have you arrested for just wandering into a situation in which you don't belong. Most good folk will simply help you out, point you to where you've told them you need to go, and send you on your way.

Again, it is important to maintain plausible deniability through it all. "What? You need one of them security cards to get into the air traffic control tower? I met a guy who said to go up and they'd show me around." It's better to be dead wrong than cold busted.

If caught, play the outsider. Any half-way intelligent security person you meet will know almost immediately that you don't belong. You can explain it away by freely admitting to being an outsider. You are supposed to be meeting a friend. You are looking for the personnel office. You don't know where the hell you are. You are looking for a nonexistent office, company, or person.

Use your credentials. If you got it, flaunt it. If you are a member of the press, a doctor, a lawyer, an inspector, a card carrying member of the ACLU, or some other kind of important person, flash those credentials when you get nabbed. It just might get you out of a jam. "I'm a photographer on assignment. Here's my press card."

Ask questions first. If you can see you are going to get questioned, beat them to the punch. Before they can question you, question *them*. Ask them how to get to room 247, or how to get to the front office, or how to get to the bathroom. You want to be more gutsy and more obvious than they expect to clear you of any suspicion. Make it seem like you're *grateful* that they've shown up to help you. "Oh, God," you say, "how *do* you get to the computer center? I was supposed to meet someone for a tour 10 minutes ago!" Don't give people a chance to set their mental wheels in motion about you. Start charming them before they start forming hypotheses about why you are there.

Don't talk too much and don't give too many details. Let people drag your story out of you. Give just enough detail to communicate what you are doing there and want you want. People have very little patience and will hear only the first few seconds of what you say anyway. You don't want to prattle off your entire story to the first person who questions you. Don't appear over-eager and anxious.

Let people do nothing. Take advantage of people's desire to do as little as possible. Provide them a convenient explanation for your presence, before they have to come up with one themselves. Offer to do all the work for them if you get stopped. "They should be expecting me! Who do I talk to about this?" Then simply step out to make the call and disappear.

Putting It Into Practice

Start small. Practice makes perfect. Try sneaking into a movie at the CinePlex 2000 before you take on the Pentagon. Start where the consequences of getting caught are insignificant.

Do some research. Know how likely it is you'll be stopped and questioned. Is security higher than average at this organization? Is it so small or tight-knit that *any* outsider is likely to be questioned? In a small company, a person found wandering about would get the familiar query, "Can I help you?" In a large company, not everybody knows each other and important visitors are more frequent. So, if a trespasser was wearing a nice suit, she'd be untouchable.

Rent a movie and watch the pros. Check out the way the Mission Impossible team smoothly gets into and out of touchy situations. Watch Artemis Gordon don disguises in *Wild Wild West*. When James Bond isn't shooting his way out of difficult situations, he is stealthing around. You can take a reverse lesson from Maxwell Smart, who demonstrates how *not* to lie: "Would you believe... I'm the plumber?"

We have everything we think we need, which isn't much. It isn't like we are going to do it on the outside. We're going to do it on the inside. All the way. We were going all the way with Susan. And she is going to show us how.

We meet her at her office at 5 p.m. to climb the tallest building in town. Philip darts around with an imaginary gun drawn, making like a 70s cop drama. The Mission Impossible tune plays in my head. Susan presses the button for the top floor, but it won't take. We've missed our window of opportunity. She darts downstairs.

In a few moments, she bursts out of the elevator, motioning us to follow. She used her connections, bluffing one of the security guys. She flicks the button for the top floor, and we hurtle upward.

We head through an unmarked door and up a ladder. We snake up another ladder and through a door that reads "Fan Room, Authorized Personnel Only." We emerge in a fluorescent lit room filled with huge metal ventilation ducts and humming bright pink, orange, and green machinery.

We clamber up another ladder and out a trap door. What I see takes my breath away. We're at the top. We're the highest people in the city on the tip of the tallest building in town.

TOUSSAINT LOUVERTURE

WAS BEING REALLY NICE TO NAPOLEON WHEN HE COMPARED HIMSELF TO HIM

by Miriam Rosenberg Roček

Quick, how many 18th century rebellions in the New World led to permanent independence for a colony from their European founders?

If you're like me, and there's no reason you should be, you live in the result of one of them; it's called the United States of America. Breaking away from Britain certainly didn't end imperialism in America, or, and this will be important, slavery; but still, being a colony that breaks away from your parent country is not an easy thing. There was only one other in the Americas: it was in Haiti, and it wins the revolution contest because, unlike the American Revolution, it not only achieved independence, it eliminated slavery. Boom. American Revolution: 0, Haitian Revolution: 1. Score one for the Francophones.

Now that that's cleared up, let's talk about the guy who, if you did a side-by-side comparison of the two revolutions, you'd probably have to call the Haitian George Washington, but I won't, because frankly this guy deserves better than to be compared to a wooden-toothed slave owner. This man is Toussaint Louverture, and he kicked *all* the asses. Not some of the asses, not most of the asses, all of the asses. In fact, in regards to the whole "what other leaders could you compare him to" thing, he personally claimed to be, basically, the black Napoleon (well, or he called Napoleon the white Toussaint Louverture. It's a little ambiguous, we'll get to that), but, as you can see from the title of this piece, that clearly does Toussaint a disservice. He was way too politically and militarily savvy to be compared to anyone who would try something as droolingly stupid as attempting to invade Russia.

No, Toussaint was not Washington, and he was not Napoleon. He was what would happen if the French Revolution had sex with Machiavelli's *The Prince* and then raised their baby in slavery with the allegorical figures of Justice and Liberty as its lesbian foster mothers and finally pitted it against the governments of all available major nations.

I'm not sure any of the above is biologically plausible, but if it is, that's where Toussaint Louverture came from.

On a non-allegorical level, Toussaint's birth was a little less promising. He was born a slave, in what we now call Haiti, in the French colony of Saint-Domingue. This was sometime around 1743. It was probably the first of November, as astute speakers of French may have gathered from his name, which means "All Saints" which is a holiday, which is on the first of November.

Anyway, where were we? Oh yeah, Toussaint was getting himself born. Some people say that Toussaint's father was a man named Gaou-Ginou, an African chieftain or king of the Arada tribe in what is now Benin. Thing is, there's really not very much reason to think that, and Toussaint actually identified another man as his father, a slave named Pierre Baptiste Simon. Of course, believing him to be the son of a king would feed into a lot of people's ideas about fate, and royal blood, and a lot of other stuff I find to be bullshit, which is why I prefer the "his father was just some guy" theory, especially as it has the added benefit of being probably true.

Royal blood or no, Toussaint had a fairly uneventful youth. In summary, he was a slave, but seems to have been relatively well treated. He even got an education, though it's hard to be exactly sure where and how. Part of what historians who are interested in him do is speculate as to where he picked up his learning, and just what he was reading; he said, wrote, and did things that made it clear he'd read a range of thinkers from the Classical period to the 18th century, and he spoke French, in addition to the Creole language spoken by Haitian slaves. He also had some medical knowledge. It's possible he was trained by some Jesuits who were hanging out at the time; he was definitely pretty into Catholicism. It doesn't really matter where he got his knowledge, what matters is that by the time he reached adulthood he knew more than enough to be dangerous. It's possible no one taught him anything at all; it's possible he simply walked up to the allegorical figure of Knowledge and talked her into telling him everything she knew.

Toussaint was set free at the age of 33. He married, and got a plantation of his own. He owned one slave in the course of his life, whom he did eventually set free. So, one day, Toussaint is hanging out, he has a wife and kids, stuff is going well-ish, when a revolution starts. I know, I know, that really sounds like the beginning of a movie starring someone like Mel Gibson. Actually, the thought of Toussaint Louverture being in the same room as Mel Gibson is so delightful I'm just going to sit here and think about it for a second. Hee, it's funny because Mel Gibson ends up violently dead. Anyway.

Toussaint may have been involved in the planning stages of this revolution, he may not. Ah, the trickiness of history. Anyway, it started with an incident known as the Boukman Rebellion, which began with a secret voodou ceremony in the woods, where slaves were called upon to rise up against their masters. It didn't come out of nowhere—this rebellion had been brewing in Haiti for a long-ass time, which is what happens when you have a system of slavery in a relatively isolated location where slaves outnumber owners ten to one. More than that, though, the French Revolution had just gone down, and slaves, not to mention the vast population of free people of color in French-controlled Haiti had damn good reason to start wondering when all that liberté, fraternité, égalité stuff they were hearing about was going to make it across the ocean. When it failed to do so (trans-Atlantic navigation was tricky in the 18th century! It probably got lost, got stuck in the Gulf Stream, and wound up in Canada), a rebellion was really the only logical step.

Here's the funny thing. It wasn't just the black Haitians who were talking about revolution. The white slaveowners were talking about having an entirely different revolution, specifically because they were afraid that all that liberté, fraternité, égalité stuff was going to start messing up their lives in Haiti! So, with the slaveowners thinking of breaking away from France, and the slaves and free blacks thinking about becoming equal, it was pretty much time for violence. And hey, the slaves had been victims of violence for a while, so it was their turn. The rebellion started with an uprising of slaves, who killed their masters and burned the plantations (the backbone of the world's sugar industry). Off to a good start!

Toussaint, when the rebellion started, actually went back to the plantation of his former owner. Not to take revenge, but to protect the family who used to own him and give them a chance to flee the attacks. You could say he had Stockholm Syndrome, or you could just say he was an amazingly good person, and objectively a better person and a better Christian (if that's something you want to get into) than any of the slave owners on the island. After demonstrating that he was an excellent candidate for sainthood (which was good, since he was already All Saints), Toussaint joined the rebel slaves in the mountains.

When Toussaint first joined the rebellion, it was as a doctor. How cool is that? He started out as a freaking combat medic. And then became a military commander. How often does that happen? Rarely. That's how often. Basically, him and Che Guevara.

France sent reinforcements, and victory was at first looking really unlikely for the rebels. At this point, they were asking for better conditions for slaves, not the actual elimination of slavery. That seemed like too crazy a goal, at that point.

Okay, this is the part where I fess up to the fact that I am not a military historian. Or an historian at all, but that's beside the point, which is that I am not at my best when talking about military strategy and the like. I am going to attempt to do justice to Toussaint's actions, but remember, this ain't my area.

So, in 1791 Toussaint was involved in a hostage standoff. Though he was holding hostages, the French ended up telling Toussaint that he and his demands—which were for better working conditions for slaves, and a bit of a prisoner exchange, not the abolition of slavery or anything crazy like that—could go fuck themselves. In doing so, they seem only to have been setting up an opportunity for Toussaint to show off how great a guy he was.

When someone doesn't give in to your demands, and you're holding hostages, you kill the hostages. That's how hostages *work*. Toussaint, though, decided that he was too good a guy for that, and saved the white hostages, even going against some other military leaders who were like "dude, this is what hostages are for." Toussaint returned them, and tried to use that as an opportunity to meet with the white leaders and talk to them like grownups. That didn't work, because they were still being assholes, but at least the hostages ended up alive, so win there, I guess, for humanity in general, and Toussaint's tendency not to kill people it wouldn't actually help him in any way to kill in particular. This might have made Toussaint appear weak, had he not gone on to pretty much kick the faces off (military term) anyone who tried to oppose him until the black Haitians ultimately controlled the whole colony.

To get there, Toussaint did a bunch of military things, in a military fashion. He was allied with the Spanish early on, because France was being dickish about slavery, and the French were beginning to recognize him as a legit leader. He was known for keeping a disciplined, European-style military camp, with soldiers trained in both the Old World style of war as well as guerrilla tactics. It was around this time that he started calling himself Louverture, which is the French word for an opening. He probably got the nickname due to his gap-toothed smile, but it says something about his awesomeness that a lot of the historians (and contemporaries) who talked about it figured it must have something to do with his ability to create openings in his enemies' lines. Man couldn't even give himself a slightly self-deprecating nickname without everyone rushing to assume it was something about how awesome he was.

Oh, by the way, that gap-toothed smile was courtesy, not of genetics or a lack of orthodontia, but of a bullet hitting Toussaint in the fucking face. Toussaint Louverture, catching bullets with his teeth. What've *you* done lately? Yeah, that's what I fucking thought. Stand up and salute when you think about Toussaint Louverture!

A couple things also happened around this time. The first big one is that Toussaint apparently decided that, fuck it, the abolition of slavery was now something he wanted to be fighting for. Aiding him in his fight against the French (and also the British) was the Yellow Fever, which killed the hell out of white people in the more tropical regions of the New World.

Eventually, France changed its mind about slavery. This was, as you are probably aware, a rather tumultuous period in French history, and it would've been weird if it hadn't changed its mind about slavery at *some* point, considering it was changing its mind about issues as fundamental as who should and who shouldn't retain ownership of their heads. The thing is, that as much as maintaining their neckal integrity was a pressing issue for French leaders, hanging onto Haiti was also a concern, and the best way to do that was to get in good with Toussaint, and the best way to do that was to change their minds about slavery. So France became pro-abolition, and, abruptly, Toussaint became pro-French.

There were a bunch of pesky Spanish and British still running around Haiti at this point, and now that Toussaint was on the French side (or, more accurately, now that France was on Toussaint's side), he got down to work getting rid of them. At one point he won seven battles in as many days. Again, I am forced to ask you what *you've* done lately. Personally, in the last seven days, I've managed to do, but not put away, my laundry. Toussaint became governor of the colony, too, just because hey, who else were they going to put in charge? Someone who *wasn't* Toussaint Louverture?

Actually, there was someone else, and this is where Toussaint's really Machiavellian nature starts to make itself clear. There was this dude, right, a white guy named Léger-Félicité Sonthonax, who had come from France to ensure equality for free people of color, and slavery for slaves. He'd succeeded at the first, but as it became increasingly clear that he wasn't going to succeed at the second, he backed down and decided he too was okay with abolition, as it looked like the only way of maintaining a hold on Haiti. It was he who, in 1793, declared the emancipation of all slaves in Haiti (the impetus for Toussaint joining the French, remember?). The problem was, that when Toussaint became governor, Sonthonax still had a lot of power, as did a French general named Étienne Maynaud Bizefranc de Lavaux. So, what's a brilliant military commander to do when faced with a couple white guys who seem to think they're in charge? Organize a military campaign and kill the everloving shit out of them, right?

Toussaint opted for something way more subtle, nonviolent, insulting, and awesome. Which, now that I think about it, was actually super French of him. He embarked on a complex political campaign against the two men, ultimately basically getting them elected to positions that would entail them going the hell back to France. Yes, he got them voted off the island. (Wow, I just dated myself with *that* reference, didn't I? I don't care, it was too appropriate.) When it comes to devious, Machiavellian methods of getting people off your back, arranging for them to be elected to a post that will force them halfway across the world wins you a billion points.

It worked on Lavaux, but Sonthonax tried to hold on to his power in Haiti, ultimately shooting himself in the proverbial foot by letting French privateers operate against American ships. Doesn't sound like something Toussaint would've cared about, except for the fact that black Haitians were trying to trade with the United States, and privateers were making that hard. In addition to fucking up trade, it gave Toussaint the excuse to finally make a military move against Sonthonax, who ended up getting the hell out of Haiti, and following Lavaux to France.

Occasionally, some asshole would decide that they were up to the task of Machievelliing Toussaint (yes, it's a verb. Because I said so, that's why.). They were so, so wrong. Two such assholes were a Brit named Maitland, and a French guy named Hedouville, both of whom attempted clever manipulations of the political situation in revolutionary Haiti. Toussaint schooled the crap out of both of them, working with one until it no longer became convenient, then spreading nasty rumors about the other to spark a popular uprising against him. You really shouldn't fuck with Toussaint Louverture.

Stuff was changing again in Europe, and now Napoleon was the boss of France. He started making new laws. Meanwhile, by total non-coincidence, Toussaint was drafting a constitution for Haiti. It had some great provisions in it like banning slavery, and making Toussaint in charge for life. It was also extremely likely to piss of France, since the whole point of it was to get in front of Napoleon's plans to make some new laws for the colonies in the New World. Not being an idiot, when it came time to present this constitution to France, Toussaint handed that job off to an underling, just to see what would happen.

Sending an underling to accomplish a task likely to piss off Napoleon is not a nice thing to do, in any sense. But it sure is smart, because that underling ended up in exile for a while, on the island of Elba (it was a popular exile spot). Basically,

Napoleon was pretty sure that this was all just a sneaky attempt on the part of Toussaint and the other black Haitians to get their island to become independent of France. Toussaint tried his best to convince Napoleon that this wasn't the case, most awesomely by subtly trying to draw a comparison between himself and Napoleon. When I say "subtly" I mean that he wrote Napoleon a letter with the heading "from the First of the Blacks to the First of the Whites." Like I said, it's hard to say whether he was saying that Napoleon was like a white Toussaint Louverture, or that he was himself a black Napoleon. Either way, it's a ballsy comparison to make when you're talking to a guy who was, at the time, considered the greatest living military genius. I'm also not sure who died and made Toussaint King of All the Black People, but I'm sure if someone had had a problem with it they would have said.

Ballsy though the letter might have been, Napoleon did not deign to answer it. This is because he was kind of a dick. It is also because he was under intense pressure to get Haiti back under French control; it was still a huge part of the global sugar industry, after all, and even if Toussaint kept insisting that he was going to keep the place nominally French, getting rid of slavery was going to have a big impact on profits, and Napoleon couldn't have that, now could he? Not when he had extremely well-thought-out military campaigns in Russia to plan.

Napoleon ended up sending some men to Haiti to restore French control, through diplomatic means. Well, I say diplomatic means. That is, *he* said diplomatic means, but for some reason he felt the need to back up that diplomacy with twenty-thousand troops. You know, for extra diplomacy. Nothing says "diplomacy" like an invading army. It is possible, I think, that the United States may be taking a few too many foreign policy tips from Napoleon. Speaking of which, the concept of a black republic was scaring America just as much as it was scaring France. There was a sense, in both countries, that the mere existence of an independent Haiti would mean violent slave uprisings, and the end of the institution of slavery itself. Napoleon named stopping "the march of the blacks" as his primary goal, while slave-owners in the United States tried desperately to keep word of what had happened in Haiti from their slaves. They'd walk around their plantations, staging loud conversations like "have you heard the latest news from Haiti? Why yes, Rhett, I have! I hear that nothing at all is happening there, as usual, and no blacks are in power there at all, and are still totally slaves." (I made that up. But that doesn't mean it's not true.)

Toussaint, not being an idiot, was pretty cognizant of the fact that when the French said "diplomacy" and then showed up with an army, they probably weren't being entirely sincere. He was fully prepared for this to become a war. It was the French who made the first aggressive move, attacking a fort along the coast. Unfortunately for Toussaint, he was right in the middle of a little bit of infighting with a rebellious general, and besides, this was the Napoleonic army we were talking about. His plan to basically let them have the coast, retreat into the highly defensible mountains, and wait for yellow fever to kill them might have actually worked, but there were some breakdowns of communication (and loyalty) that led to some of the generals not making the retreat. In the end, Toussaint, and his family, were captured by the French forces.

Toussaint was sent to France as a prisoner. On his way, he made his most famous and most bad-ass comment ever, informing his captors that by removing him from power they had cut down "only the trunk of the tree of liberty; it will spring up again from the roots, for they are many and they are deep."

He turned out to be completely right about that. Although he died in prison less than a year later (pneumonia... kind of the yellow fever of Europe, when it came to killing people from different climates, though the conditions he was held in certainly didn't help), Haiti would go on to achieve complete independence and lasting freedom that same year. It was the world's first independent black republic.

Napoleon's dickish comment on the shitty way he had treated Toussaint (years after he had been forced to give up all of his holdings in the New World, as a direct result of the success of the Haitian Revolution) was "what could the death of one wretched Negro mean to me?"

You know, Napoleon was kind of an asshole. And as disingenuous as he was racist, because for a guy he didn't care about, he sure spent a lot of time trying to kill him.

So that, in brief, is the life of Toussaint Louverture, the man who brought freedom to Haiti. His legacy has been celebrated in countless ways, and that's all well and good, but here's a thought to take away with you when you consider what gets a historical hero recognition in the modern day; Danny Glover has been trying to get a movie made about Toussaint for years now. Hollywood just isn't interested. Because, you see, there are no white heroes to bring in the crowds. Deal with that for a second. We don't get a movie about one of the greatest military minds, and greatest fighters for freedom, because all the white guys in his story act like asshats. Toussaint Louverture is being punished, post-mortem, for the fact that virtually every white guy he ever met either owned him or wanted him dead.

Man, I was going to make a final point. I forget what it was. Fuck the movie industry for believing white audiences wouldn't want to see a movie without a white hero, fuck white audiences for repeatedly proving them right, and fuck the movie industry again for thinking white audiences are the only ones that matter.

Miriam posts as Steampunk Emma Goldman on her blog anachro-anarcho.blogspot.com

Nevermind the Morlocks,

Here's Occupy Wall Street!

By David Z. Morris
Illustration by Tommy Poirier-Morissette

From the editor: I knew I wasn't the only steampunk caught up in the wave of anti-corporatist sentiment that called itself Occupy Wall Street. When I joined the Oakland general strike of November 2nd, I saw a woman wearing brass goggles to protect herself from pepperspray. Unwoman was playing cello for the crowd that was busy shutting down a corporate bank. And the Port of Oakland picket line was staffed by a pair of steampunks on stilts. Beyond that, the internet-versed were disseminating the information and videos that gave the movement its lifeblood. I put a call out for folks interested in writing about Occupy and steampunk, and I'm quite happy with David's response:

Since its inception, *SteamPunk Magazine* has staked and defended the claim that steampunk—as a culture, as an idea, as a movement—is inherently political. This position was run particularly far up the flagpole in issue #7, with Margaret Killjoy's essay on steampunk and politics and Jaymee Goh's on steampunk and race. Those essays were part of what made me so excited when a piece of my fiction was accepted for publication in this, issue #8. It felt like I'd lucked into a community that shared both my aesthetics and my values.

Just as the writers and editors were beginning to get the current issue in fighting shape, politics went from the page to the streets as Occupy Wall Street somersaulted into the public consciousness. Working to highlight economic inequality and organized on anarchist principles of horizontality, mutual aid, and consensus, the form and substance of Occupy Wall Street struck a nerve deeper than even its initial organizers had hoped, and local groups popped up across North America and Europe. Many involved with *SteamPunk Magazine* (including myself and our esteemed editor) found themselves absorbed in the unfolding movement. We can hardly help, then, but reopen some of the questions addressed only one issue ago (if nothing else, we have to justify our dawdling). Was the broad interest of *SteamPunk Magazine* contributors in Occupy just happenstance, a random overlap of unrelated concerns? Or is there a deeper connection between this particular strain of fantastic fiction and the politics of Occupy?

At the risk of retreading a well-worn path, we can start off by echoing Margaret's point that most if not all of the core writers and works of

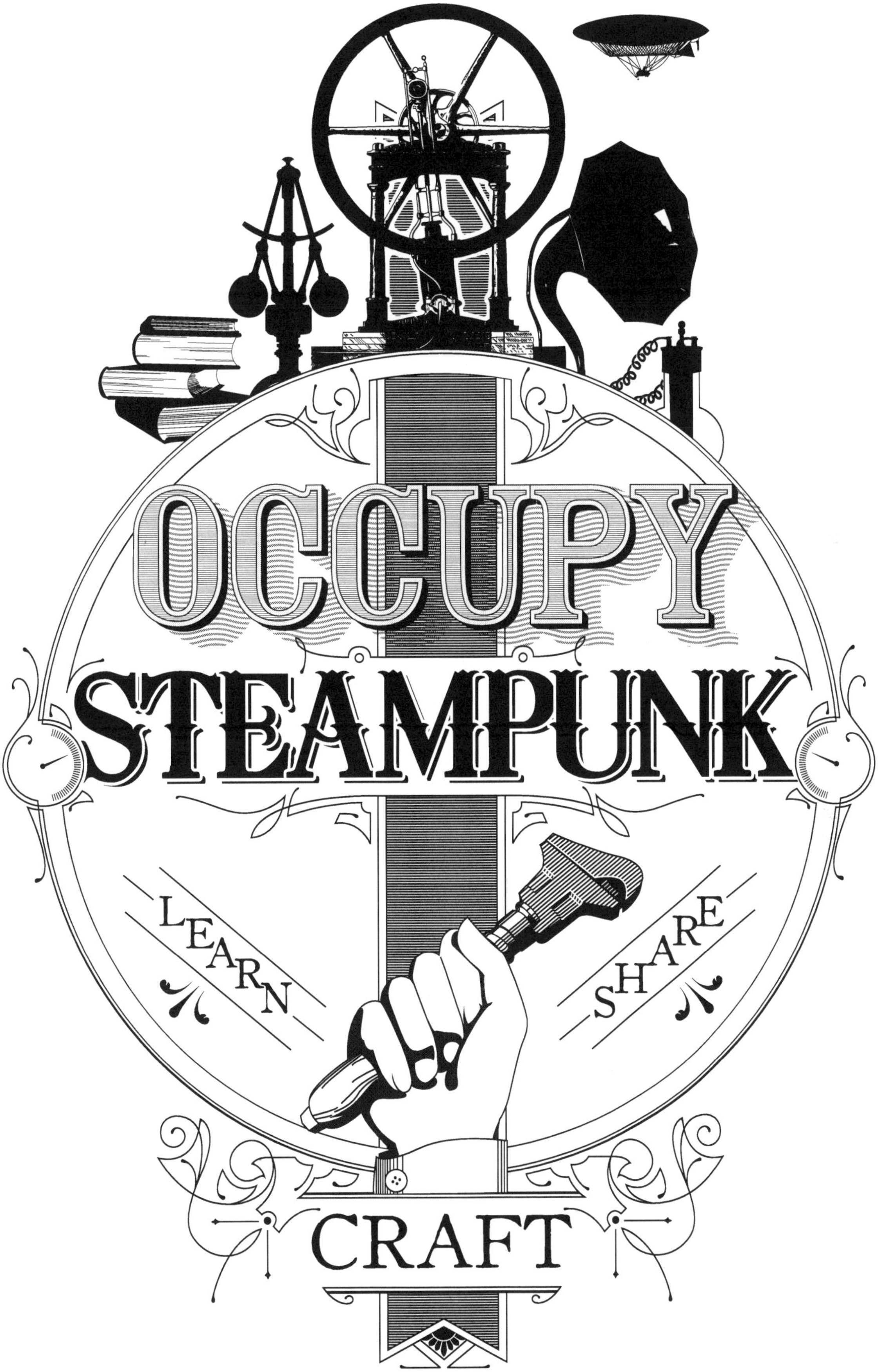
OCCUPY
STEAMPUNK
LEARN
SHARE
CRAFT

both vintage, "roots" steampunk and contemporary "steampunk proper" have been explicitly populist, anti-hierarchical, and anti-imperial. Ms. Killjoy provided a great rundown of the anarchist or socialist values of Jules Verne, H.G. Wells, Michael Moorcock, and Alan Moore, so I'll just add Mary Shelley's *Frankenstein,* China Mieville's New Crobuzon novels, and M. John Harrison's Viriconium cycle. Victor Frankenstein implicitly robs pauper's graves to construct his creature, and Mieville's Remade are the product of a penal system that relies on convict labor, their awkward steam prosthetics a symbol of everything from Victorian debt peonage to the contemporary prison-industrial complex. Harrison's work, set in the bleak leavings of a collapsed techno-utopia, is one of the harshest, most grinding condemnations of our own foolhardy greed anywhere in literature. From the core concern with income inequality to the corollary issues of biopower and ecology, then, the "steampunk canon" resonates deeply with the concerns of Occupy.

But Occupy isn't just a set of issues, and neither is steampunk—they're both also methods, structures, and toolboxes. To unpack this we have to emphasize that, when you ask whether something "is political," you're actually asking two questions. You may be asking whether it has political *significance*, or you may be asking whether it has a particular political *agenda.* Steampunk, in the first sense, is political to the bone, because in all its variations it is concerned with the relationship between technology and society, and specifically, how technology affects power relations. The genre (as a subset/masterset of science fiction) was born during, and remains grounded in, the Industrial Revolution, which upended a huge variety of social norms incredibly quickly—in the most obvious instance, by redistributing power from the landed gentry to capitalist entrepreneurs. The reshaping of society by technology and other social forces hasn't stopped, and it's still often hard to know just where we're heading. That's why steampunk is appealing: like all science fiction, it's about giving us settings, images, and ideas through which we can metaphorically debate the political issues that we as a society are confronting.

Steampunk gives us a space for discussion that also happens to be thrilling and imaginative, and Occupy has proven so compelling because it does exactly the same thing. Occupy's core concept of appropriating public space to have ongoing conversations about issues is much like Michael Moorcock's decision in *The Warlord of the Air* to appropriate the cliches of swashbuckling space opera for a sweeping condemnation of imperialism. What had once been a place for light fun and games is injected with a dose of deadly seriousness—without losing the sense that just about anything is possible.

I'm active in Occupy in the southeast US, a region where neither the traditional Left nor activism are very strong. But with its invitational, open-ended structure, Occupy has provided a "big tent" for people with some very different viewpoints, and even for those still trying to figure out what their politics are. Rather than shutting down anyone who you disagree with on some minor point, the structure and spirit of Occupy encourage you to find common ground. In particular, southern Occupations have proven damn near irresistible to supporters of right-libertarian Ron Paul. Occupy, with its focus on space and conversation rather than just shouting and sign-waving, has been a chance to discover the common ground between these libertarians, traditional leftists, and anarcho-syndicalists (surprise—there's a lot!). Occupy has already shifted the US national dialogue from debt ceilings to income inequality, but this connection of a wide array of perspectives may be the most profound lasting impact of Occupy.

This may be a good way to think of those steampunk fans who object to this magazine's political stance, claiming that the genre is just about having fun. Margaret's essay last issue made the fine point that this crowd's rejection of politics is in fact a politics of its own, that tophattery and zeppelin fetishism romanticize the Victorian elite and obscure the social structures that gave such privilege to a few dashing, clean-cut adventurers. But maybe it's more productive to approach the gogglerati as our very own Ron Paul contingent—despite some serious differences, there are plenty of signs of common ground.

What all steampunk fans have in common, whatever their attitude to politics, is a sense of wonder and a fascination with unfettered possibility. This is why the technology at its heart is essentially malleable, limitless, and so frequently mixed with magic—steampunk is definitely "science fantasy," and the fantasy that fuels it is the possibility of a world radically different from our own. The elitist, romantic, neo-Victorian wing of steampunk is concerned (more or less metaphorically) with the potential for industrial capitalism to produce fantastic weapons, while anarcho-socialist-politico-whatever steampunk imagines those weapons turned against the masters of the system. But at least on the face of it, both are fascinated with with change, with innovation—with *really cool stuff.*

When you ask whether something "is political," you're actually asking two questions. You may be asking whether it has political *significance*, or you may be asking whether it has a particular political *agenda*. Steampunk, in the first sense, is political to the bone.

Technology has a very tricky relationship with power. Especially if we're talking about old-school industrial technology, it requires the concentration of wealth to produce but then radically empowers an individual to work outside of the constrictions of the everyday. You can see the resulting tension in Jules Verne's Captain Nemo: while Nemo is a raging anticolonialist, he's also an Indian noble who has only been able to realize his dream of radical individual freedom by extracting wealth from peasant laborers and building a sweet submarine. And while he laments the deaths of the enemy sailors he "must" kill, he's still going around terminating a lot of other people's freedoms in his quest to defend his own. The debate over whether steampunk is "political" is, in substance, a debate over how we should feel about technology, and about progress—is it great, or is it grim? Do the means of hierarchy justify the end of turning (some small set of) people into minor gods?

Another way of saying this is that, because of steampunk's centering in an imagined alternative past, it can be read in two directions. For the "isn't technology rad" cadre, a 19th-century raygun is a metaphor for the potential of our own society to continue advancing into new frontiers, technological and otherwise. For the more skeptical among us, Mieville's Remade are a lament for where we went wrong, a chance to think of how things might have gone differently, and just maybe, how we can backtrack to a time before we screwed it all up. In practice, these aren't so much diametrically opposed viewpoints as they are two sides of a dialogue about where we go from here.

The most important thread uniting Occupy and steampunk is also a very, very basic one. Both are driven by the idea that *another world is possible.* Occupy feels, in the best possible way, like starting from square one—groups of like-minded people have come together and built organizations from nothing, working to figure out the most basic elements of what they're doing. They've ditched the baggage that has hemmed in most US activists for the past forty years: just as steampunk has never been beholden to any practical limits on how technology "actually" works, Occupy isn't willing to be hemmed in by the way politics are "supposed" to work. If the pragmatic, territorial, and dour Thomas Edison reflects the status quo of electoral politics, Occupy has much in common with (the not necessarily historically accurate popular image of) steampunk's patron saint, Nikola Tesla. Both have their heads in the clouds, their gaze directed to something that is half goal and half dream. Both go beyond what makes sense to grasp at new and wild visions, imagination and passion driving them from the practical into the magical.

Of course, as many reading this will note, Tesla came to a solitary, impoverished, and obscure end, while Edison went on to enduring wealth, power, and near-deification. Tesla's audacity and singular passion led him to overreach, and to neglect some of the humdrum practicalities of modern professional life. Edison might have been tagged the "Wizard of Menlo Park," but he was much less a magician than a businessman. Maybe we would live in a better (or at least more interesting) world if Tesla had been better at negotiating existing systems like trademark and investment… but can you square that circle? Is it possible to remain a true visionary while becoming more and more implicated in the system you're trying to see beyond? That's exactly the challenge confronting Occupy—and one where steampunk can offer not just intriguing resonances and fellow-feeling, but a rich conversation, already long in progress.

In the Shadows of Giants

by David Redford

Embarkation Tower 2 at Skye Edge Landing was a modest structure as these things went, since the port was not built to handle the very largest ships. Still, in the opinion of most who spurned the lift and its threepenny fare, the hundred and fifty steps to the spectators' platform provided more than enough exercise. From here you could see most of Sheffield, if you cared to: from the gigantic steel mills and coal tips of the East End to the tower of the town hall, and to the hilltops of Wincobank, Pitsmoor, Shirecliffe and Crookes, each with its own light-tower burning 24 hours a day.

Paul Kennedy gripped the handrail and steadied himself, breathed deeply, and with a great effort of will let go of the polished rail with his right hand. He focused on the forward lounge of the *Hibernia* and waved, even smiled weakly. With a bit of luck she wouldn't notice the awkwardness of movement, the fixed grin that had taken the place of—what? Some great Shakespearean oration, a declaration of his undying love in front of her husband and children? No, that wouldn't do at all. Even someone of his humble origins knew not to cause a scene—or more accurately, knew when and when not to cause one—and besides, he was never one for flowery language.

"Don't go," you see, that said all it needed to say; not that it had done him any good. He tore his gaze away from the privileged cargo and took a better look at the instrument of his misfortune. A thousand feet long, they said, and from here he could believe it, every inch. Nothing to scale them against when they're floating a mile or so up, although he remembered the games he played as a kid on Scotland Street, holding his breath for the time it took the immense shadow to pass over. Took some doing, especially when the sun was getting low.

A harsh cry from the Landing Master brought him back to the present, and he watched the small army of boys rush to do the man's bidding, loosening off the inch-thick stablising lines one by one. They worked their way round the catwalk, letting the deceptively frail-looking lines fall before shinning back down to earth along one of the four anchor ropes that remained. Not a job for the faint-hearted, or the sluggish for that matter; the stevedores were already yelling at the slowest of the balloon monkeys, shifting their 21-pound leather-faced hammers, making as if to strike the peg out early and send the 3-inch rope barreling back into the bowels of the ship on its spring-loaded capstan, small boy and all. But the boys knew it was an act as much as the hammermen did; it was a matter of professional pride to strike all four pegs at the same moment, and they would no more strike one early than light a cigarette.

Still, grabbing a rope was not a bad way to get yourself to New York, Paul mused, if you could avoid getting wound round the capstan and didn't mind being frozen to death and starved of oxygen for two days straight. There were some that had tried, if you believed the stories the anchormen told. And places where you sometimes gave the stew a miss if one of the big transatlantics had just docked. Kennedy had laughed at this; he thought he knew a tall story when he heard one, but figured "why take the chance?" His diet was about varied enough as it was.

Superstition, that was all; the same superstition the hammermen had about the odd stray spark flying 50 feet across the void and setting the hydrogen off. Everyone who knew anything about airships knew this was impossible, but still…

Paul was suddenly aware that one of them had struck up a song, and the others soon joined in a rousing chorus. But he was not listening to the words, which the ground crew took pride in peppering with as many obscene references as possible; he was scanning the crowd of passengers who had come to the rail to get their vicarious experience of the Dignity of Labour, laugh indulgently, and then leave this filthy place behind at a mile a minute. It came back to him then with a jolt; she really was going, she was going now, and he couldn't see her for this damn crowd.

But there was her bowler-hatted mediocrity of a husband—something financial, he guessed—and there were the two boys, the elder explaining the words of the song to the fascinated glee of the younger—but where was Sarah?

And then he saw her, on the far side of the lounge, looking straight out the front of the gondola as if the factory chimneys of West Bar and Neepsend held the utmost fascination for her. As if she didn't dare face him, Paul thought—but just at that moment her jaw seemed to set and she turned, her fan coming up in front of her face and moving to and fro in a seemingly random pattern.

He was dimly aware that the stevedores' bawdy song had come to an end, that four great hammers had swung in perfect unison, that four steel pegs had flown from their oak sockets and clattered across the dock, that the great ship was already starting to rise and turn; of the realisation that Sarah's shoulders appeared to be shaking; and, shortly before she blurred from his vision, of what her message had meant.

Forget.

When he looked up again the *Hibernia* was already over a mile away, heading west to Dublin for the connection to New York; but still it loomed large, as if taunting him over what he had lost. A train was gone in a minute or two, Paul reflected, but an airship could stay in sight for half an hour. He wasn't going to put himself through that.

He turned away, descended the stairs of Skye Edge Landing and headed back towards town, the labyrinthine and lawless courtyards of Duke Street either side of him. He could hear the half-starved kids running to and fro, playing, fighting and picking pockets. Not so different from the streets he had grown up on; the ones Sarah was flying over right now.

Kennedy found himself thinking of the kids on Scotland Street, not so long ago, and tried to remember what it had been like, playing in the shadows of the giants.

On Lighter Than Air Craft

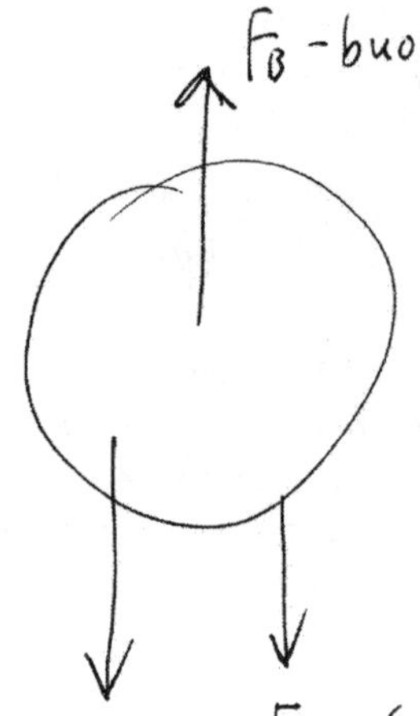

F_B - buoyant force - equal in magnitude to force of gravity on air displaced by the aircraft.

F_ℓ - force of gravity on lifting gas

F_c - force of gravity on aircraft structure.

when just floating, $F_{net} = 0$

$\therefore F_B = F_\ell + F_c$

$m_B g = m_\ell g + m_c g$

(m_B ← mass of air displaced; m_ℓ ← mass of lifting gas in balloon; m_c ← mass of craft)

but $m = \rho V$, where ρ is density, V is volume.

$\therefore \rho_B V = \rho_\ell V + m_c$, V = volume of balloon(s)

$\therefore m_c = V(\rho_B - \rho_\ell)$

↳ to float $\boxed{m_c \leq V(\rho_B - \rho_\ell)}$ ①

The lift of your balloon depends on the difference between the density of air, ρ_B and your lifting gas, ρ_ℓ; the mass of your craft must be less than the value given by $V(\rho_B - \rho_\ell)$ in order to float.

How much lift?

$F_{net} = F_B - F_\ell - F_c$

$F_{net} = (\rho_B V - \rho_\ell V - m_c)\,g$ ②

lift mass $= \rho_B V - \rho_\ell V - \underline{m_c}$ *

The mass of the craft is very much under your control in design (as is volume of the balloon itself).

Values for ρ_B and ρ_ℓ can be found in tables, or calculated for any temperature or pressure from first principles.

ON LIGHTER-THAN-AIR CRAFT

by P. Fobbington

To fly the stately airship again, ah, even to be but a passenger on such a vessel is a dream of many and represents flight in its true sense. No need to buzz loudly through the air just to stay aloft, one can float as they like. By some calculations an airship with a 1000 lb lift capacity would cost less than $5000 to build, but before one starts producing their own completely unlicensed and half-airworthy craft, one needs a little physics on their side to guide the design.

For lighter than air craft, no matter whether they use hot air or lifting gases such as hydrogen (or helium, that unexciting, expensive gas in limited availability), the mathematics all comes down to how the density of your lifting gas stacks up against the density of air. The principle of buoyancy holds sway here and it is worth stating formulaically that "The upwards force on an object suspended in a medium (air, here) is equal to the weight of the volume of medium displaced by the object."

In the derivation given, two equations are presented—the first shows what maximum mass you can have for the structure of your airship's frame, including the material for the balloon, and passengers for a given lifting gas and volume of balloon. The second equation, just a rephrasing of the first, will tell you for a given mass of your craft how much lift you will get.

It is easy to see that if the density of the lifting gas were 0 (a vacuum balloon!… sadly not practical), you would have maximum lift. On the other hand, if your lifting gas was the exact same air as outside the balloon, your lift capacity would be equal to the mass of your balloon—downwards!—and you would obviously go nowhere. Tables are available on the internet and in chemistry and physics textbooks for the densities of various gases under different, commonly encountered temperatures and pressures. Please take a look at the example given, where the result calculated is that a cubic meter of hydrogen gas will lift about 1.2 kg of payload.

Hopefully in times to come we can discuss the derivation of an equation that will let you find the density of any gas (or known mixture of gases) under any temperature and pressure conditions (so long as you have a periodic table handy), the effect of water vapour on gaseous densities, how to design lighter than air craft with lifting gases at higher working temperatures (i.e. hot air balloons… or hot hydrogen balloons!) and ultimately some explorations into more interesting unexplored designs.

At 0°C and 101.325kPa ambient air pressure,
the density of air is $\rho_B = 1.2920\,g/L$,
the density of hydrogen is $\rho_e = 0.08988\,g/L$
— g/L (grams per Litre)

For a balloon with a volume of 1000L (imagine a cube 1m on each side)

EXAMPLE

$$m_c \leq V(\rho_B - \rho_e)$$

$$m_c \leq 1000L\,(1.2920\,g/L - 0.08988\,g/L)$$

$$\therefore m_c \lesssim 1200g \Rightarrow 1.2\,kg$$

∴ A 1000L balloon filled with hydrogen under these conditions could lift 1.2 kg, including the mass of the bag to hold the hydrogen!

Air For Fire

A Steampunk Hypatia

by David Major

HYPATIA (b. 350–370 – d. March 415 A.D.) was a Neo-Platonist philosopher and scientist, and a leading academic and teacher in Alexandria, in the Roman province of Egypt. Her students came from all over the known world, from both inside and outside the Roman Empire. She admitted both Christians and Pagans to her classes—she apparently observed no distinction between the two—as well as both men and women.

Alexandria was one of the main centres of early Christianity, while still having a large and active Pagan community. Hypatia became involved in a political dispute between Cyril, the Christian Patriach of the city, and Orestes, the Roman Prefect, who was a Pagan.

Her death was ordered by Cyril (now Saint Cyril, a Catholic "Church Father") in 415 A.D.

A Christian mob attacked her outside her house and dragged her to a church. There, they flayed her with seashells and pottery shards, then dragged her through the streets before dismembering and burning her. At which stage of this torture she actually died will never be known, but one account describes her "convulsing limbs being torn from her flayed body."

THE FLOOD WATERS RECEDED, AND THOSE WHO HAD SURvived by taking to the mountains or the tops of the highest hills returned to the valleys and to the plains, and there they found that everything was covered with silt and mud and debris, and they sighed, and set to work.

And even though the heat of the New Sun would sometimes dry the land until it cracked, and other times the rain would come so heavily that the crops were washed away, life went on. Somewhere, animals and seed had survived, and what was needed for building was found, and so, although life was harder than it had been before the change, and it was uncertain and too often full of woe, it did go on.

But that was all a very long time ago. Things settled down. Today, we have order.

"Cardinal Synesius,"—I have been asked more than once, in so many words—"you were close to the pagan Hypatia, one of her students—what really happened? And how is it that you have become a Cardinal?"

Here is what happened concerning the pagan Hypatia.

"Librarian..." I am watching the approaching Walkers from the main window of the Soil Feeder Archive. They clank, wheezing steam and smoke, through the corn fields, the foliage thrust aside by their iron flanks, rows crushed beneath their riveted hooves, their crews unaware or uncaring of the damage they are doing to the fragile crops.

I hear her sigh. She has come to stand beside me. "What do they want now? Are we expecting anything?" She says it to herself; there is no use asking me.

Librarian Hypatia is... well, one of them. Half born, half made. She is a relic of the old world; we don't grow them like her any more, and we certainly don't make them like her any more—not that we could; even if we wanted to. We don't stand as straight or tall, we don't think as fast, and we don't remake ourselves the way they are built to. They are from the past, a past not like anything you or I have ever known. Most people these days don't talk about the ones like her. Me? Oh, I want to know...

Hypatia has been the Head Librarian at the Serapeum since her predecessor became something to do with the Cardinals (apparently an offer he could not refuse). He taught her well; she knows where every Soil Conditioner, every Enhancer, every Charm and Relic, every Amulet, every Air, Karma and Water Filter, and every book and scroll is kept. The entire contents of the Serapeum Library are catalogued in her mind. And I get the feeling that she understands it all, as well.

When the Walkers bring in shipments from the Cardinals' factories, she knows where each should go, and when the time comes to send stock out to the Markets, she knows exactly what is needed and where. Without her, the Markets would barely work. There would be chaos.

And of course I love her, beyond all reason; but I have never told her that, and that is something I regret, because now I never will. So please, say nothing, to anyone.

The Walkers are Machinist Tippit's. We know him.

"New stock," he says, taking a glass tube of documents from a pouch in his suit and handing it to me. He won't hand it to Hypatia; he never does.

We have never seen Machinist Tippit's face. We have never seen what hides behind the stitched leather mask, or the obsidian darkness of the eyeshieldings; even his gait is hidden beneath the folds and volume of the hermetically sealed suit he wears.

I open the tube.

"Fire Feeders," he says, before I have a chance to begin reading. "There have been new developments with fire." His voice has taken on a new tone; hubris, I think it is called. I do not like it.

Workers, strapped into clanking and hissing lifters, are already unloading crates of the new devices from the Walkers' holds.

"All instances of fire will now require one of these. All instances."—A brief pause. "New developments with fire," he repeats.

Hypatia reaches into a crate and retrieves a cylinder of tinted glass, apparently empty, and mounted in a framework of brass piping and tubes, from which extends an array of knobs, keys, and valves.

There is no sound as she moves, her tattooed skin slides easily among the prosthetics. Her machinery moves with practiced, easy precision. Her eyepiece clicks softly as she studies the device.

"Machinist Tippit, on whose authori–"

"The Council of Cardinals," he interrupts her. The Machinist hides neither his impatience nor his disdain. He does not like the Serapeum, and he does not like us. "Every instance of fire, whether industrial or domestic, must be fed the output from Valves Two, Four, and Subletting Valve D, which must be engaged concomitantly with the Release Control Activator and the Filter Engagement at Locus Two. We..."—he gestures with a gloved hand, in such a way that it is clear that he is referring to himself in particular— "...will be adding Fire Infringements to the list of Surveillable Activities and Warrantable Exigencies. In short—librarian—every fire will require a Fire Feeder." He pauses, for a rasping breath. "And your... establishment, of course, will distribute them."

"I see. New developments with fire, indeed." Hypatia hands the device to him, smiles as he hesitates before reaching to take it. Even through his mask and suit, his discomfort at the possibility of physical contact with her is palpable. "No doubt this has all been well thought out. Good day, Machinist." She turns to me. "Come, Synesius."

As she turns to leave, Hypatia gives me the look that I have seen before, the one that means —there is more to say, and —but not here, and —oh, who are these clowns? and —be careful… all somehow rolled into one.

You might think that she could be staid, or too studious (do they mean the same thing? I would have to ask her, that's the sort of thing she knows without having to open a book or a scroll), or perhaps she might be quiet, or even dour. She is a Head Librarian, after all, so none of these would be a surprise, would they?

But none of this is the case. She has a raw, untrammeled, this-is-what-it-is intelligence. Like a zebra, she cannot be tamed. She applies her mind to whatever she chooses, with an enthusiasm that, I know, some see as wanton. She disturbs some people, makes them uneasy.

She does this: she traces and records the paths that the birds draw in the sky and the creening of the fish in the streams at night, and the height of the Silver Kros flower stalks after three days of rain, and the rhythm of cloven hooves, which when enumerated, makes her heart thrill—and she says that there are so many ideas and facts, and things that might just be facts, that sometimes she feels as though her mind is going to explode in a wonderful kaleidoscope of thoughts, and each thought is a flame, or a spark, and it's like a huge, endless orgasm, of stars, and thinking, and wonder, and gratitude, and love for it all, and amazement that it can be so deep and clear, and full of light, and steal your breath without trying at all, or really doing anything, and she laughs and says that's what thinking is, and that's what science is, and what love is, and it's all one thing; and there's no difference between any of it.

So. I can't argue, can I? I just say yes.

"I do not trust the Machinists, Synesius."

I know. I say nothing. Pathetically, I hope that my silence will encourage hers. Things are complicated enough already; but she continues.

"And I do not like these Fire Feeder things. There is something wrong."

Of that, I am not so sure. Their reasons may be obscure at times, but the rule of the Cardinals gives us order. They keep us all fed, they keep the Markets full. They have allowed the Library to exist; the Theological Office pays many of our expenses. And the people have their Amulets and their Soil Conditioners and their Filters—all at the behest of the Cardinals. So yes, I am unsure.

"I am going to look into this." Hypatia's eyes ripple with anticipation. She writes a note on her arm, on the square of bare skin—often used for this purpose—just beside the hasp of her left elbow assembly. When she writes there, it is always something that she will get to very soon.

I finally accept that I must say something.

"I will be busy, cataloging the soil samples from the farms," I say. "Perhaps I will help you later," I add, intending, but failing, to lie.

I go to the Market sometimes, when I need time and space to think. I know there's some sort of paradox there. The Market is crowded, full of noise and chaos; and—despite being at the foot of the cliff face that the Library dominates and therefore close enough for the two to be considered parts of a whole—it always feels like a world apart.

Everything in the Library feels so important, but in the Market there are so many things to do without, and it all feels so unimportant. Oh… it's not worth getting complicated about. I come here to think.

They have been selling the Fire Feeders for a month now. A stern-faced Fiscal Probity Marshall from the District Planning College has given the stallholders who qualify the requisite paperwork and certificates; he has collected the Onsellers' Fees and License Application Bonds, and has seen that the Market Committee's Undertaking to Accept Responsibilities Charter has been amended, cited, ratified (twice) and signed. Pending approval by three Inspectors from the Planning College and countersigning by the Theological Office, everything is in order. The stallholders will soon be able to make formal applications.

As always, Croesus the Lydian is there, with his piles of Magnesium Amulets (for the pig farmers) and Fertility Harmonic Essentials (for the croppers, and for domestic use); and his collection of Conditioners and Filters, of course. Since the Planning College ended the sale of real food at the Markets, Traders like Croesus have done well.

Masked though he is, I recognise him by his modifications. Beneath them he wears the crudely articulated but well-sealed suit of a Trader, not all that different from the suits worn by Farmers and Labourers. The supple leather outer skin and pneumatic tubing protects him from the air, and the heaviest eyeshieldings available protect his sight from constant exposure to the glare of the New Sun. All going well, he will not be blind for some years yet.

"Synesius! Look at this." Croesus holds up one of the new Feeders.

"Yes, I know. I've seen them."

Croesus grunts. "These things… Look what they're telling us…"

He hands me a sheet. Yes, I've seen this before, as well. But I stand there, and I read it again. The Cardinals have taken scientific advice to the effect that raw fire is destructive to the new air; that for the sake of the common good, and so that the air is not consumed and so that all life on the earth does not perish, if not very soon then at least soon, and if not soon then within a few short generations, and do you not care about your children, and what sort of world are we going to leave for future generations?... and so on, then on account of all those things, and for the common good, they, the Council of Cardinals, decree that all fire must be accompanied by a Fire Feeder, and that any case of fire being unattended by a Fire Feeder is most deleterious in its effect on the common good, and will be dealt with appropriately, which is to say, harshly. All this said, the sheet finishes with the admonition that the Machinists will police the new regulations vigorously.

I had forgotten about all this. Or I had tried. Hypatia has seemed preoccupied for the last week, and I know it has something to with these accursed cylinders. Her work has suffered, and her mood has changed. The notes that she has been writing on her clothes and skin have been taking on a different tone, one that I haven't seen before, and on several occasions, she has even hidden them from my sight.

I sigh. I know that Croesus has something for me. "Alright, my friend. Tell me..."

It is easier for him to take me to his village.

"This," he says, as he shows me an iron smelter with furnaces that are cold and silent. "They cannot afford the fees, nor the Scribe to complete the forms."

"And this," as he opens the door of a Farmer's hut, and someone hurries to hide an illegal flame. "They do not qualify."

"And here." A seed winnower has left a week's work behind, to go to prison because of unpaid debt. His yard is in disarray. "His family will wait for a time, and then they will despair, and then they will go, and the last of the seed will disappear into the wind."

"Tell him, Croesus," the villagers say, when they see the insignia of the Library on me. "Show him the ruined food, the empty purses, animals and children gone hungry for fear of the Walkers who come stalking fearsomely out of the mist, how we struggle to share one Fire Feeder between too many hearths... We were poor before, and now look at us... At least we have our Faith, and our Amulets..."

I have seen enough. "These things are the ruin of these people," I say.

"You have it, my friend," Croesus replies. "There was a time I would happily have made money by selling toys to these people, but no more, Synesius. I have done this for long enough."

When I return, Hypatia is in her laboratory. Something I almost recognise is being dismantled, sprawled across a table, a collapsed cage of tubes, clockwork, and instruments.

"Librarian. I will help you."

She barely looks up, but I do see a smile. "At last. Now, take this. And this. Dismantle them. See whether they match. Whether they can serve a common purpose, or whether they are specific to the tasks which they apparently perform. And then put them back together. Please."

And soon I know what she has been doing, and that she has been right all along.

She has explained everything. It is among her best work, I am sure. The rest of us post copies of it on doors and walls in the Villages, hand them out at the Market below the Serapeum and at all the other Markets—as far as we can reach. Soon everyone knows what she has discovered; that the Fire Feeders do nothing. They are useless, inert, their parts are not internally connected in any way that serves a function. Everything about them and of them and in them is for show and does precisely nothing.

Hypatia draws the proof, she describes it in words and pictures, and supports the case using anecdotes, comparisons, metaphors and similes. The point can escape no one; the argument is undeniable.

The word spreads through the Markets, and beyond, to the Villages and Farms, to the other Libraries, to the large Cities... Soon, the news has travelled everywhere, and soon after that, no one is buying Fire Feeders any more.

Crowds gather, and in anger they burn their Certificates and Permissions and Notices of Allocation on fires lit, pointedly and loudly, without the aid of Fire Feeders. Feeders are broken and trampled, and thrown at the officers from the Planning College, and then at their Walkers as they retreat, shamed, unratified, and penniless.

The Priests from the Theological Office stand by and watch, conferring in low, troubled voices.

I am concerned. "This could get out of control," I say to Hypatia.

She smiles. "A free mind cannot be commanded, Synesius. Let the people discover their freedom, it is time that these traces were pulled at..."

But I can see that Hypatia has noticed the men from the Theological Office, and I have noticed them, and they have noticed us. The air is thick with noticing.

"How pleasant, another visit,"—but Machinist Tippit is having none of such niceties.

"You are responsible, and you will stop it," he rasps. His anger! You can hear his outrage seething behind his mask.

It is a tense moment; he has not come alone.

Whatever his intention, it doesn't work. The pleading (disguised); the threats and blustering (not so disguised). None of it works. He even tries—"... and they nurse their ignorance, Librarian; they've grown to like their sores so much that they scratch them with their dirty nails to keep them festering. Words are no use. Only the force and violence of being ruled over absolutely, to the core of their being, makes any sense to them..."

"Free minds, Machinist," says Hypatia, looking out the window at the circle of the Moon in the sky, as though that in itself is enough, and he will understand.

He says nothing, but my fear is that he does understand, and that if he does not, then his Cardinals will.

So. Machinist Tippit's report to the Cardinals must have been delivered.

There is violence now. The supporters of the Cardinals have appeared, and they swarm like black ants, and there seem to be so many. They burn and strike and attack, and they are aided by Machinists who pour out of Walkers everywhere...

They are at the gates of the Library, and now they have pushed their way through, and people are wounded and calling for help, but there is none to be had; there is fighting, and look, there are flames, and now there are dead, and the statues of the Gods are falling and crumbling... Serapis, save yourself...

There is fighting everywhere now. She knows what I am thinking.

"Everything flowers, Synesius. All entelechies unfold, according to their natures. Yours. Mine. Theirs. That is all there is."

The gears in her arms whir as she passes the works of the Philosopher to me. "Get these to safety, Synesius," she says, her voice clicking.

The roar of the crowd draws closer. She seems to notice that her skin and clothes are covered with writing. "And this must go," she says, and turns towards her rooms.

Buildings burn through the night. The Serapeum's books and scrolls lie in drifts of flame and smoke and ashes. Machinists trample through the fires, unfeeling of the heat, pouring fuel on the flames.

In just a few hours, a thousand years is dust and ash.

The next morning, Walkers beyond counting stream out of the early morning gloom into the towns and villages, through the smoking ruins of the Library and what remains of the Market, coughing clouds of searing gas, loudspeakers blaring:

—The air is in danger of dying. Renegades have been burning fires without permission and without the requisite devices in attendance. We are spraying remediant to counter the imminent danger of environmental collapse. Despite the irritation to your lungs, please breathe as deeply as possible. What sort of world do you want to leave for coming generations?

The sky darkens. The sun is hidden behind a mass that suddenly hangs over everything. I have seen engravings of what is above us, so I know what it is. But still...

It is a bolo-airship, so large that its prow and stern seem to disappear into the haze of the distance.

The Cardinals are here.

There are many of us, herded into a hall aboard the bolo. I have been taken too, but I am not in chains; I am not being tried before the Council of Cardinals. It is even whispered that if I am careful, I have a future.

Hypatia was among the first taken. She stands chained, upright, and she meets their gaze without faltering. I doubt that I have ever loved her as much as I do right now. What follows achieves nothing.

—So you are the wheel, are you? Hypatia says to them, and they do not understand, but I do, because I have seen the butterfly tattooed on the curve of her back, just where I like to think that she shivered with delight when I breathed upon it, lightly, on the last of the three occasions that I was close to her.

—I think that the ocean is deep enough, she says later, and again they do not understand, but again I do, and I am not surprised that she says nothing more to them after that, and that there are no tears when they command the Machinists to come and tie her to the rack on which she will now surely die.

There is some sort of ceremony. They pronounce something, observe some sort of outer form, I don't know. I can barely hear. Is this vengeance, or a lesson for the masses? The hall is crowded.

I realise now that I have been waiting for this moment, anticipating it so much that I feel an awful relief now that it is finally here. From now on, everything that moves, everything that is thought—everything—they all link up, the pieces roar like the wind as they come together, like the cogs in a machine that has no way of stopping.

The inevitability of it shudders in my mind.

The names of the things that I see begin to dissolve; the performance I watch now is beyond names. The Cardinals gather around, and insist that I stand with them. I would look at the floor, but I cannot.

Machinists, Tippit among them, with knives and saws and screwdrivers and tools—too small and delicate to determine their names or uses with any certainty—surround Hypatia, and begin to dismantle her.

"I suspected by the look of him that he enjoys killing things," she says to me, looking up at Tippit as he removes her arms, unscrewing and cutting, cursing, but not impatiently, as her blood soaks into his instruments. She bites her lip, her eyes close for a second, and then open again.

"Does it hurt?" I ask, hopelessly. I can think of nothing else to say.

"The pain is breath-taking," she replies, as a saw begins to open her chest. "Space and time, they are mere irritants, Synesius,"—but it is harder to understand her now, because her mouth fills with bloody foam as she speaks.

A Machinist begins to remove wire from her chest cavity.

"Aah, the heart…!", and the Cardinals gather around. "Let us see the heart!" and there is glee among them.

"What can I do?" I whisper.

And there is a glance that will cut me in half forever, and then the light in her eyes has gone out.

And I cannot see Hypatia, and I cannot recognise the things and pieces that were her, nor the Machinists, who are still cutting and pulling and excising, nor the Cardinals, who are full of joy and happiness—they are all just parts, and pieces, and their names have all gone.

And as these things fall away, dripping into the dust where her blood gathers—there I can almost recognise what I thought to be me, the thing that gave names to the world, and I see now, quite clearly, that there is nothing there.

The flesh which three times I touched is gone. It has been ground with seashells and fed to the dogs.

In the end, the Cardinals were full of joy at the death.

They had been surprised that there was no heart to be found inside her. There were organs, true, and blood, true, and cogs and gears and devices of exotic and wonderful manufacture—but there was no heart.

They wondered at this, but only briefly. Then they ordered the flesh disposed of, and the mechanical parts cleaned of blood and gore, and sent to the factories, to be used there, assimilated into the machines that make the Amulets and Feeders and Filters and Conditioners.

"This will make things easier," said the Cardinals. "Our production values will be greatly enhanced. Going forward, we expect positive impacts on Liquidity and Market Penetration. Stability has returned!"

But I know where her heart is; I think I have known all along. Her heart was in the ease and effortless power of her machinery, it was in the heat and pulse of her flesh, and the way the rhythms of them both flowed from the one to the other, from the other to the one, in the way it all found form in her thinking, and science, and love—there was never any difference between any of them…

And every idea that she had, the urging of every thought towards the freedom of mind and spirit—they all left their mark, an imprint of her intent, her joy, in every cog, every gear, every piece of brass and obsidian; they were all her; there was no difference between any of them…

And now, do you see it?

She is everywhere, taken and installed in the machinery of their factories, where the production lines create one thing after another, charm after charm, device after device, again and again without end, every one with Hypatia in them, imago, complete and perfect, imprinted not just in their form but in their essence, and from there they go out to the world, delivered by Hordes of Walkers to Markets and Libraries everywhere, where one day—yes, one day, there will be enough.

So, there you have it. There is your answer. That is what happened concerning the pagan Hypatia. But please, say nothing of this.

REVIEWS

We are happy to consider published and performed matter for review, including books, albums, plays, movies, and the like. Written work needs to be physically published, but we consider albums that exist only in electronic format. We are not interested in reviewing websites. To submit material for review, query us at COLLECTIVE@STEAMPUNKMAGAZINE.COM

FICTION

THE DOCTOR AND THE KID: A WEIRD WEST TALE

MIKE RESNICK

Pyr

Reviewed by S. Kimery

A steampunk tale of the Weird West kind, *The Doctor and the Kid* is Mike Resnick's sequel to last year's *The Buntline Special*.

The year is 1882 and the westward expansion of the United States has been blocked at the Mississippi River by the powerful magic of native medicine men like Geronimo. Thomas Edison has been dispatched by the US government to battle them with Science, aided by intrepid engineer Ned Buntline. Doc Holliday, fleeing his notoriety (and enemies) after the gunfight at the O.K. Corral and wasted by tuberculosis, tries to impress the touring Oscar Wilde with his poker skills and ends up losing a fortune. Faced with financial ruin, Holliday decides to try his hand as a bounty hunter, and the biggest payoff out there is a new gunslinger called "Billy the Kid." Drama and intrigue ensues.

Mike Resnick is a science fiction legend. If he had carved a notch in his six-shooters every time he was nominated for a Hugo (5 wins out of 35 nominations and counting) there wouldn't be much left of them to hold on to. So, with an author who has demonstrated his mastery of the craft over hundreds of short stories and tens of novels (so far) it is at least slightly disappointing that *The Doctor and the Kid* doesn't give its readers much to hold on to.

In the opening chapters I was left wondering "Oh boy! Oscar Wilde is drinking with Doc Holliday: how can *this* go wrong?" Well, it's not that so much goes wrong with the book, there's just so little that goes very right.

Both Wilde and Susan B. Anthony are introduced at the get-go, but they are more dropped names than actual characters. Both just wander out of the tale within a few pages. Edison, we are told at least twice, a hundred pages apart, has an amazing steampunk mechanical arm that can do all kinds of cool stuff... but it isn't put to use in the story. "Chekhov's gun" has its trigger lock firmly in place, with many ingredients of a substantial tale left as loose ends in this entertaining page-turner.

After a point I stopped noticing what *wasn't* in the book and started becoming aware of increasing numbers of "repeat offenders." Do you remember how in *The Da Vinci Code*, Dan Brown introduced virtually every female character as "sporting an attractive v-neck that hinted at ample cleavage"? Resnick's Doc Holliday keeps reminding everyone in earshot of his classical education and that he is absolutely going to die soon, and has several discussions with the featureless Edison/Buntline duo ending with minor variations of "I hadn't thought of that, but I'm just a dentist that drinks too much!"

The thin veneer of steampunk elements are perfunctory yet unimaginative. The Bunt Line's (get it?) electric armored coaches are stagecoaches without horses, brass robot saloon gals are creepy robot saloon gals (bring your own lube), and some buildings are impervious hardened brass buildings! Even the gosh-wow weaponry that does make an appearance is there for just an instant before the story sheds them like a prom dress.

The Doctor and the Kid captures the spirit of a period western dime novel, albeit with some gears glued to the cover. Despite its shortcomings, it would make an excellent light read for a long flight or stage coach ride, but you'll feel few qualms about leaving it on the seat for the next passenger to enjoy.

NON-FICTION

STEAMPUNK GEAR, GADGETS AND GISMOS: A MAKER'S GUIDE TO CREATING MODERN ARTIFACTS

THOMAS WILLEFORD

McGraw-Hill/TAB Electronics

Reviewed by Prof. Offlogic

The above styled book (hereafter referenced as *SGGG*) is a very, very useful guide to creating one's own steampunk props. Accessible to the gamut of fabricators, from Master (and Mistress) Contraptors to beginning makers, Willeford offers not only his considerable insights as to achieving "that certain look and feel," but his own perspective on the "steampunk aesthetic" (at considerable risk!).

SGGG is a rich resource, featuring not only an introduction by Kaja (and artwork by Phil) Foglio (the masterminds behind *Girl Genius*), but photographs, patterns and philosophical musings.

The highly controversial "What is Steampunk?" section is sure to inflame message boards!

How-to topics like "Tools of the Modern Mad Scientist," "Gear Mining—How to Dissect a Cuckoo," and "The Art and Philosophy of Scavenging" are followed by similarly illustrated/photographic chapter-by-chapter guides to producing your own "Aetheric Ray Deflector Solid Brass Goggles," "Calibrated Indicator Gauges," and "Professor Grimmelore's Mark I Superior Replacement Arm and Integrated Gatling Gun Attachment" among other highly useful equipment.

Willeford's thoughtful and very hands-on guide to producing your own steampunk "gear" is well thought out and splendidly presented. The photographs and diagrams are most illustrative and detailed. The recommendations regarding "flea market budget" finds and general maker-craft is just the sort of valuable information that can convince you that you can do it yourself your own way.

MUSIC

CARNIVAL SYMPOSIUM

SPIKY

Reviewed by Cassandra Marshall

What started out as a richly layered and cyclical sound quickly divided me. On one hand, there's this gorgeous industrial sound like in the slow beginnings of "The First Cog Lament" and the intense dreaminess of "The Last Cog Lament" that quickly became a favorite, but things go downhill as soon as anyone starts singing.

The album features Captain John Sprocket from the American steampunk folk band The Cog Is Dead and Jessica Donati from the French symphonic metal band Ivalys and while I'm sure they must sound well enough separately, together it's like someone threw a wrench in the works.

The dear Captain has some interesting inflections to his voice at times, but the cheese grater sounds he "sings" overshadow the lyrics so much that it's a good thing the digipack comes with a lyrics booklet (which is atmospherically illustrated by Aurelien Police). You can even hear the poor chap cough halfway through "Dancing on a Fence." Donati usually sounds well enough but her parts are so few and just about always in competition with the Captain. Let's just pretend the twelve-minute mess "The Chronophagist" doesn't exist.

As for the lyrics, they seem just about as confused as those singing: Empty barrels make the most noise./ There's a coffee's cup somewhere,/In need of a feeling./ Real tears worth more than fake smiles. And then later on, I could see my hatred every day/To make me feel human, If I could say/O Please, I'm trapped within me/ And after all, I'm crying coffee.

Me? I'm just crying for my eardrums. Maybe if we all beg, they'll release an instrumental-only version, one that would surely make their inspirations Hans Zimmer and Danny Elfman proud. Spiky could take a lesson from Zimmer and Elfman; they don't need noise to blow you away.

MUSIC

THESE BUBBLES COME FROM ANTS

HELLBLINKI SEXTET

Reviewed by Margaret Killjoy

I've been a fan of Hellblinki Sextet since I first saw them play in 2005. If you took a gothy, dark, one-man band from New Orleans and exploded him out into a full group, complete with accordions and musical toys, you might have Hellblinki. But as much as I've loved seeing them perform, their recordings have usually left me wanting. There are tracks I return to time and time again from their previous releases, such as the amazing rendition of "Bella Ciao" (an Italian anti-fascist partisan song) on *Oratory*, but the albums themselves didn't seem to hold up. I feel that way no longer. *These Bubbles Come From Ants* nails their sound and their genius.

The male vocals are grizzled and charming; the female vocals operatic and resplendent. The songs are an equal mix of creepy and beautiful.

In many ways, Hellblinki sums up what I want out of a steampunk band. They aren't "steampunk" themed. When steampunk adopted them into its fold, they didn't start hot-gluing gears to their instruments. They just kept doing what they were doing: playing their strange brand of dark, experimental American folk with a snake oil salesman aesthetic.

Standout songs on the album include the blues/ rock'n'roll of "Don't Go Down To The Woods Today" and the atmospheric "Ants For Now."

Let the Banal Fall by the Wayside

by Dimitri Markotin

"Our reality is not their reality even if our fantasy sometimes looks similar."
—Libby Bulloff

A punk friend of mine from Spain, a marine biologist, recently came to work in a lab in the States for a few months. The language barrier was rough on her, but the culture shock was worse. "I don't know what to do," she told me, in so many words. "There's this punk boy who is trying to date me. He says he respects me, and he respects my politics, but some of the punk he listens to is rightwing and he's friends with nazi skins too."

Sad-hearted, I tried to explain to her that in the US, being punk means almost nothing besides fashion.

"It's hard for me to understand," she told me. "In Spain it's not like this."

She told him she wasn't interested, and I'm sure he thought she was close-minded for rejecting him just because he listened to rightwing music. I doubt he realized that it was his own non-addressing of bigotry that drove her away.

It's fair, and depressing, to say that punk in the US is pretty much just a scene these days. To quote Shakespeare for no real reason, "It is a tale told by an idiot, full of sound and fury, signifying nothing." In much of the world, allegiance to punk means something. For all our faults, we punks stand in opposition to the world we feel beats down upon us and others. But while there are thousands of us political punks in the US, we're lost in a sea of scenesters concerned only with aesthetics, people who ostracize everyone who isn't punk enough, who don't wear the right clothes or keep up with the right bands. My friends and I are degraded as "PC punks" because giving a shit isn't cool anymore. I've even heard, from people I can only presume to be ignorant of punk's roots, that anarchists and anti-fascists are co-opting the scene. Or that we're close-minded for not putting up with nazis.

Does this sound familiar?

The steampunk scene is not the same thing as the punk scene. You don't have to be a punk to be a steampunk. You don't need to have ripped clothes or a nose ring or a mohawk or patches or tattoos to look the part. I'm not equating these two things—I'm just drawing parallels.

Steampunk developed as a branch of literature deeply critical of imperial culture and the assumptions made by those who idealized a certain kind of progress that was taken for granted during the Victorian and Edwardian eras. Steampunk pitted the genius inventor not against the "savages" of the colonized lands as period literature did, but against the colonizing forces. Much as cyberpunk forgot about spaceship captains and made the everyperson the protagonist of their own lives, steampunk brought the urchin to the foreground, and it gave prostitutes the knives and Gatling guns they needed to strike back at a society that had mistreated them.

And this will always be part of what steampunk is to me. It's an aesthetic lens with which to see and experience the world. It's dress-up, yes, but it's not *just* dress-up.

Call me a PC steampunk all you want, but there are thousands of us. We're not all anarchists. We don't all organize against fascism, or organize at all. We weren't all involved in Occupy. Hell, some of us might not have even been inspired by Occupy. We're not the "real punk steampunks." We're not even the "real steampunks." There's no label for us besides "steampunk," because we don't need one.

Maybe we're drawn to steampunk for the whimsy too. Maybe we like playing make-believe as much as anyone. But fun is more fun when it challenges us, and I would argue it's even more fun when it's challenging and changing the world.

Sometimes I feel like my Spanish friend. My jaw drops with culture shock when I hear people earnestly defend empire in the comments section on a blog and call themselves steampunk. And like my Spanish friend, I'm afraid that if I call them on it, they'll see *me* as the bigot.

By and large, this whole rant is nothing but bluster. Because the steampunk scene is actually pretty cool. The people I meet at steampunk conventions are excited about steampunk because they're excited about artisan culture and they're excited about forging a community in the face of the unimaginative society that surrounds us. They're excited about challenging, fun fiction, and they let the banal stuff fall by the wayside where it belongs. There are problems, but we're addressing them.

"Punk" to me is a compliment. Punk doesn't mean "has a mohawk," punk means "thinks outside the box, searches for culture and meaning in their own way." I want "steampunk" to be shorthand for "awesome," not shorthand for "wears goggles on their hat." And I may not get my way, but fuck it, I'm going to try.

SUBMIT TO NO MASTER!

BUT CONSIDER CONTRIBUTING TO STEAMPUNK MAGAZINE!

send submissions and queries to: READERS@STEAMPUNKMAGAZINE.COM

Please keep in mind before submitting that we publish under Creative Commons licensing, which means that people will be free to reproduce and alter your work for noncommercial purposes. We are not currently a paying market. Please introduce yourself in your introduction letter: we like to know that we're working with actual people.

FICTION: We appreciate well-written, grammatically consistent fiction. That said, we are more interested in representing the underclasses and the exploited, rather than the exploiters. We have no interest in misogynistic or racist work. We will work with fiction of nearly any length, although works longer than six thousand words will be less likely to be accepted, as they may have to be split over multiple issues. We will always check with you before any changes are made to your work. Submissions can be in .RTF, .DOC, .DOCX, or .ODT format attached to email. Please include a wordcount of your story in your email. We recieve more fiction submissions than the rest of the categories combined and reject a majority of what we receive.

POETRY: We are not currently accepting poetry submissions.

ILLUSTRATION: We print the magazine in black and white, and attempt to keep illustrations as reproducible as possible. Ideally, you will contact us, including a link to your work, and we will add you to our list of interested illustrators. Any submissions need to be of high resolution (300dpi or higher), and preferably in .TIFF format.

PHOTOGRAPHY: We do not currently run photography.

HOW-TOS: We are always looking for people who have mad scientist skills to share. We are interested in nearly every form of DIY, although engineering, crafts, and fashion are particularly dear to us.

COMICS: We would love to run more comics. Contact us!

REVIEWS: We run reviews of books, movies, zines, music, etc. We are looking to expand our reviews section and will no longer just be running reviews of work we find exceptional. We are looking for reviewers as well: please contact us if you are interested.

FASHION: Although we are quite interested in steampunk fashion, we are more interested by DIY skill-sharing than exhibition of existing work. If you want to share patterns or tips for clothing, hair, or accessories, then please let us know!

INTERVIEWS: Most of the reviews we run are conducted by staff of the magazine, but we do occasionally accept interviews conducted by others. However, it would be best to contact us before going through the work of conducting the interview to see if we would be interested in running an interview with your subject.

ADVERTISEMENTS: We do not run paid advertisements on our site or in our magazine. We rarely run press releases. If you have a product you want our readers to be aware of, your best bet is to write a how-to article explaining how to make it, submit the media for review, or request an interview.

OTHER: Surprise us! We're nicer people than we sound!

SteamPunk Magazine is published by Combustion Books, a worker-run genre fiction publisher based out of New York City. Issue #8 was released in January, 2012.

EDITOR: Margaret Killjoy

ART EDITOR: Juan Navarro

CONTRIBUTING EDITORS: Libby Bulloff; J Boone Dryden; Carolyn Doughtery; reginazabo;

CONTRIBUTING WRITERS: Larry Amyett, Jr.; Cassandra Marshall; Profesor Calamity; Katherine Casey; The Catastrophone Orchestra; Mikael Ivan Eriksson; P. Fobbington; Kate Franklin; Margaret Killjoy; E.M. Johnson; S. Kimery, David Major; Dimitri Markotin; Screaming Mathilda; Wes Modes; Marie Morgan; David Z. Morris; Jamie Murray; Juan Navarro; Profesor Offlogic; Pinche; David Redford; Miriam Rosenberg Roček; James Schafer

CONTRIBUTING ILLUSTRATORS: Manny Aguilera; Tina Black; Scary Boots; Sarah Dungan; Doctor Geof; Allison Healy; Tommy Poirier-Morissette; Juan Navarro; E.M. Johnson; Larry Nadolsky, Sergei Tuterov

FONTS: Tw Cen; Adobe Garamond Pro; P22 VICTORIAN GOTHIC; Ehmcke

Comic by Doctor Geof

11107539R00070

Made in the USA
Charleston, SC
30 January 2012